Praise for

Essential Composting Toilets

For over a hundred years our society has been treating water and waste as necessary evils rather than resources to be celebrated. A Victorian paradigm that is incredibly wasteful and damaging continues to underlie how we handle our most precious resources in nearly every city and community on the planet. There is a better way. In this important book the Baird's shed critical insights on the power and rightfulness of composting our waste and closing the loop between nutrient and fertility. Their depth of knowledge, practical experience and collected examples provide a way forward that is responsible and regenerative.

— Jason F. McLennan, CEO, McLennan Design,
and founder, the Living Building Challenge

Get your shit together the right way, on the first try. This book is one of the best researched and presented books on the topic of composting toilets that I have ever read and is a must-have for anyone wanting to create resilient closed-loop systems on their home, acreage, or farm.

— Rob Avis, P. Eng, co-author, *Essential Rainwater Harvesting*,
www.vergepermaculture.ca

As a builder and manager of several humanure composting systems I can enthusiastically recommend this essential guide! Gord and Ann Baird begin this book with a simple statement: Thermophilic composting at the right temperature for the right amount of time transforms humanure into safe and nutrient rich compost. From this essential point they present the full range of options for reconnecting the missing link in our nutrient cycle. From the home scale to the urban office building, solutions are thoroughly examined and clearly explained.

— Darrell Frey, Three Sisters Farm, author, *Bioshelter Market Garden*
and co-author, *The Food Forest Handbook*

True to form, Ann and Gord Baird once again lead on the innovation front with their informative and timely guide "Essential Composting Toilets". Their research, pragmatism and practical experience, has resulted in a significant contribution to the advancement of water conservation practice in BC.

— Eric Bonham P.Eng,
Partnership for Water Sustainability in British Columbia (PWSBC)

Gord and Ann Baird have written a totally practical, inspiring guide that shows how we can work in partnership with the bacteria and worms to build healthy, safe residential composting toilets. Their dedication to detail is astounding. All praise to the psychrophilic and mesophilic bacteria — and to the Bairds!

— Guy Dauncey, author, *Journey to the Future*

Gord and Ann take on this potentially shitty subject and present it with a wonderful balance of science and humor. As we embark upon an owner-builder off-grid home project, this book could not have come at a better time. The guidance provided for working with building officials is invaluable. It helped us ask the right questions, and gave us additional confidence that we were making the right choices.

— Jeff Walton, Cowichan Valley.

sustainable
building
essentials

essential
COMPOSTING
TOILETS

a guide to options, design,
installation, and use

Gord Baird and Ann Baird

new society
PUBLISHERS

New Society
Sustainable Building Essentials Series

Series editors
Chris Magwood and Jen Feigin

Title list

Essential Hempcrete Construction, Chris Magwood

Essential Prefab Straw Bale Construction, Chris Magwood

Essential Building Science, Jacob Deva Racusin

Essential Light Straw Clay Construction, Lydia Doleman

Essential Sustainable Home Design, Chris Magwood

Essential Cordwood Building, Rob Roy

Essential Earthbag Construction, Kelly Hart

Essential Natural Plasters, Michael Henry & Tina Therrien

Essential Composting Toilets, Gord Baird & Ann Baird

See www.newsociety.com/SBES for a complete list of new and forthcoming series titles.

THE SUSTAINABLE BUILDING ESSENTIALS SERIES covers the full range of natural and green building techniques with a focus on sustainable materials and methods and code compliance. Firmly rooted in sound building science and drawing on decades of experience, these large-format, highly illustrated manuals deliver comprehensive, practical guidance from leading experts using a well-organized step-by-step approach. Whether your interest is foundations, walls, insulation, mechanical systems, or final finishes, these unique books present the essential information on each topic including:

- Material specifications, testing, and building code references
- Plan drawings for all common applications
- Tool lists and complete installation instructions
- Finishing, maintenance, and renovation techniques
- Budgeting and labor estimates
- Additional resources

Written by the world's leading sustainable builders, designers, and engineers, these succinct, user-friendly handbooks are indispensable tools for any project where accurate and reliable information is key to success. GET THE ESSENTIALS!

Cover design by Diane McIntosh.
Interior background texture © Adobestock 90796513

Printed in Canada. First printing November 2018.

Inquiries regarding requests to reprint all or part of *Essential Composting Toilets* should be addressed to New Society Publishers at the address below. To order directly from the publishers, please call toll-free (North America) 1-800-567-6772, or order online at www.newsociety.com

Any other inquiries can be directed by mail to:
New Society Publishers

P.O. Box 189, Gabriola Island, BC V0R 1X0, Canada
(250) 247-9737

LIBRARY AND ARCHIVES CANADA CATALOGUING IN PUBLICATION

Baird, Gord, 1969-, author
        Essential composting toilets : a guide to options, design, installation, and use
/ Gord Baird and Ann Baird.

(Sustainable building essentials)
Includes bibliographical references and index.
Issued in print and electronic formats.
ISBN 978-0-86571-872-2 (softcover).--ISBN 978-1-55092-665-1 (PDF).--
ISBN 978-1-77142-260-4 (EPUB)

        1. Toilets--Design and construction. 2. Compost. 3. Sewage as fertilizer.
4. Night soil. I. Baird, Ann, 1967-, author II. Title. III. Title: Composting toilets.
IV. Series: Sustainable building essentials

        TH6498.B35 2018   696'.182   C2018-904152-8
                                                                 C2018-904153-6

New Society Publishers' mission is to publish books that contribute in fundamental ways to building an ecologically sustainable and just society, and to do so with the least possible impact on the environment, in a manner that models this vision.

Contents

Acknowledgments

As JOINT AUTHORS, we would like to thank each other for patience, listening, persevering, and dividing of tasks. Gord did the majority of the research, the initial draft, and drawings, while Ann went through and rewrote most of the text while deleting and adding content along the way and gave extensive feedback on drawings. We both take credit for the poor humor … sorry about that. The whole process involved endless discussions while having a lot of trust in each other. Writing a book together is both more difficult and much easier … thankfully, the latter was the dominant experience.

We are most grateful for the editing support of many good friends who provided feedback for us to consider and incorporate into the various draft manuscripts. Thanks to Paul Doherty, Chris Magwood, Christina Goodvin, and Jeff Walton. Your feedback was detailed, critical, hilarious, honest, and even somewhat painful to read.

This book would not exist without the immense work of Ian Ralston, who has been incredibly influential in the design of the Province of BC's Sewerage System Standard Practices. Ian was the lead author of the *Manual of Composting Toilets and Greywater Practice* for the BC Ministry of Health.

Special thanks go to Jason McLennan who we admire greatly for his inspirational and visionary leadership as the lead author of the International Living Building Challenge of which our Eco-Sense home participated. Many years ago, Jason introduced us to the concept of Net Zero Water and the role of compost toilets in regenerative design.

We also acknowledge compost toilet guru Joseph Jenkins, whose inspirational and pivotal book called *The Humanure Handbook* first inspired Ann a few years before she met Gord. In fact, for Gord to make it to their third date, Ann required Gord to read Joseph's book. Gord passed the test, and their future path was set in motion.

And finally, we bow down in awe to the real heroes of this book: the bacteria, worms, fungi, and arthropods who had this shit figured out millions of years ago.

Thank you all,
Ann and Gord

Chapter 1

Introduction

What We Cover

THIS VOLUME OF THE *SUSTAINABLE BUILDING ESSENTIALS SERIES* focuses on residential compost toilets for the North American audience. It is a comprehensive reference for selection, design, installation, management, best practices, and safety concerns. This book is for homeowner/builders, contractor/builders, architects, designers, and ecological design students. Regulators and policy makers will also find value in the content. Various compost toilet systems will be presented along with the evaluations of each system that will help the reader select specific design applications.

With home-scale compost toilet systems, the homeowner has the primary responsibility for the day-to-day use, care, functioning, and servicing of the system. For this reason, it's the homeowner who needs to choose a system with full knowledge of the implications. The information in this book is designed to ensure the owner or design professional fully understands the choices. This book references regulations and research from North America and Europe; it is technical enough to be used by regulators and policy makers, yet practical enough to be understood by homeowners, contractors, and designers.

Although we have attempted to define terminology throughout, there will be times when the reader may find it useful to reference the Glossary included near the end of the book.

The regulatory environment is changing for all building technologies. It is moving away from prescriptive building codes to ones that are objective in nature, based on guidelines that make room for the plethora of proven alternatives — as long as they meet the functions and objectives of the regulations. This book attempts to bridge the gap between regulatory jargon and regular language. After all, going to the bathroom should be a simple, natural, and safe process. No degree should be required to complete the task!

This book begins with the most basic question of whether compost toilets are suitable for you (or your client). From there, we explore the importance of regulation and the biology of composting and pathogen death, with the primary goal being safety. Next, we discuss fundamental components of systems, design considerations, and calculations on system sizing. Not every reader will need to work through all the calculations, but we felt it was important to include this section for those who desire this level of detail. With these basics under your belt, we lead you through the different types of compost toilet systems, looking at design, key considerations, and their strengths and weaknesses. This "flows" into fluid management for urine and leachate, and we lay out a step-by-step process for sizing leachate tanks and calculating soil infiltration. We finish with a brief glimpse into the paradigm shift for hi-tech toilet technology currently underway — an effort to completely reinvent the toilet. It's a development poised to disrupt the way we view waste management.

Compost toilets are often linked with greywater systems, but greywater treatment is its own very specific topic, requiring discussions about soils, dispersal methods, and science that this book does not cover.

A homeowner's choice to incorporate a compost toilet may involve dealing with health officials and regulators; understanding the science and language that they use is critical. Their job is to manage public safety.

For the regulators who read this book, we urge you to evaluate your preconceptions and assumptions surrounding compost toilets and human waste. The science that dictates how we treat human wastes in large, centralized sewerage systems is the very same science that governs the processes involved with compost toilets; in both cases, the objective is to ensure that outputs are safe and pollution is avoided.

Not Just a Rural Solution

When we began writing this book, we had the misconception that the best way to service urban populations was still the standard water-based infrastructure. Our extensive review of the scientific literature has led us to a different conclusion: compost toilets have significant applications even in suburbs and cities.

A movement away from water-based sewerage systems for cities has become a growing focus for researchers and planners. How could this occur? And why should this occur? It's not lost on scientists studying agriculture, nutrition, engineering, epidemiology, sociology, and ecology, that our present water-based sewerage systems are complicit in negatively impacting our health and the environment. This converging science is exciting, yet we are still witnessing delays in policies and regulations to keep up with that science. Biases and preconceptions are strong, and it will take many more

initiatives similar to the "Reinvent the Toilet Challenge" by the Gates Foundation (see Chapter 10) to drag Western culture into acceptance of viable alternatives.

From Waste Stream to Mainstream

So, let's get started. First of all: *There is no such thing as a composting toilet.*
You might think it odd that we would start a book about compost toilets by stating that no such thing exists. But this book challenges the idea that a toilet can compost its contents. It can't. Composting is a specific process, one that occurs under specific conditions — and those conditions do not exist in any toilet. No doubt, stating this will raise the ire of many manufacturers of "composting toilets." Manufacturers, don't despair! We share the same aims. But our intent here is to make sure homeowners and regulators understand what it takes to design a *compost toilet,* one capable of converting raw materials into a sanitized, benign material through biological means.

Compost toilets come in a wide variety of shapes and forms, from site-built systems to systems that manufacturers have invested millions of dollars to research, design, fabricate, certify, and market. Think of that the next time you have the idea that these systems are not common. These systems fill market needs throughout the world and are commonly found in modernized countries like Australia, Sweden, Finland, Norway, Germany, New Zealand, as well as a host of East and Southeast Asian countries (North America is behind on this trend). In some places, their use arises from necessity, as a result of escalating water shortages. In others, their use arises from societal values placed on resource recovery. Regardless of the motivation, all rely

---

Big problems with waste treatment and nutrient deficiency in soils could both be solved through appropriate technology and design. One solution: Composting toilets.

Composting toilets do not exist — because composting does not happen IN the toilet.

on an understanding of the science around *ecosan* (ecological sanitation).

With global population expected to rise to 8.6 billion by 2030, 9.8 billion in 2050, and 11.2 billion by 2100, increased stressors will be placed on the availability of food, clean drinking water, and enough water for agriculture. Additionally, there is a large migration from low- and middle-income countries to high-income countries. All of this is, and will continue to be, exacerbated by a climate changing so rapidly it is outpacing even the worst-case predicted scenarios (Wuebbles et al., 2017). Centralized, water-dependent waste systems will become luxuries; they will not be able to keep up to overwhelming growth. Additionally, limited availability of the nutrients required to support agriculture make it senseless to continue flushing them down the toilet (Department of Economic and Social Affairs, United Nations, 2017).

Though plant-nutrient flows should be circular, present waste-handling makes them linear; both septic systems and flushing permanently remove nutrients from the natural soil cycle. And we can't afford to lose them. Phosphorus, for example, is a critical element used in agriculture. With five countries controlling 85% of the reserves, and a dwindling supply due to over-mining, we are seeing massive price shocks — as demonstrated in 2008, when there was an 800% increase in the price of phosphorus (Cordell and White, 2014). Recovering that dwindling resource from the waste stream will soon become an economic imperative.

Compost toilets (CTs) are essentially a progressive system that collects and handles human feces and urine so that they can be safely composted. Where they already exist, CTs form part of the infrastructure used in removing compostable and biodegradable solids from a hydraulic (water-based) sewage disposal system, thus allowing the opportunity to convert the waste materials (resources) into an ecologically beneficial nutrient source in a safe and hygienic manner — that is, sanitized.

The toilets themselves do no composting: "Composting is a managed process of bio-oxidation of a solid heterogeneous organic substrate including a thermophilic phase" (Canadian Council of Ministers of the Environment and Compost Guidelines Task Group, 2005). In other words, true composting meets three conditions:

- It is managed by humans (it is rare for it to occur in nature).
- It is aerobic, requiring oxygen.
- It generates its own internal biological heat.

If these three conditions don't exist, it's not composting. Inside a compost toilet, biological decomposition processes do occur as soon as all that stuff leaves our body and becomes exposed to the air, but that is not technically composting. And extended periods of decomposition may transform materials, but true composting is a much more rapid process. We'll have a complete discussion of decomposition and composting in later chapters.

The basic aspects of using a compost toilet are straightforward: 1) You go to the bathroom. (Any questions?); 2) The deposit is collected in vessel; 3) That collection is then either minimally processed to a mature-enough state that it can be buried and thus safely reintroduced to the environment, or, better yet, it is further composted to a state that sanitizes and reduces pathogens to a level so safe it can be used as a beneficial nutrient resource.

Though we will look more closely at the concepts of maturation and sanitization later

North America needs an urgent update to our cultural belief to match the overwhelming scientific consensus on how to safely compost human manure.

Waste doesn't exist in nature. There are only *resources*.

(in Chapter 2), it is timely to introduce them here: *Mature composts* are those that have decreased nitrogen, no odor, and are safe to plants and animals; *sanitized composts* have no disease-causing organisms. Safety and best practices ensure the creation of a product that meets standards for intended use.

Compost toilets ARE NOT pit toilets or outhouses where deposits are collected in saturated anaerobic conditions that ultimately become highly unpleasant (unless you're a fly) and potentially harmful.

When we take composted or sanitized materials and reincorporate them into the environment with no negative impacts, we, in essence, do not create waste. Compost toilets are a tool for collecting and processing materials so they do not become waste.

Geographic regions in the world where water and/or agricultural soil amendments are scarce have been beneficially composting their resources for generations. Certain cultures, such as the Hunza in Pakistan, have been using human manure composting systems responsibly for thousands of years in a cycle of food production and human resource recovery. However, the collection and spreading of raw, unprocessed human manures as field fertilizer (referred to as *night soil*) — although a common practice in many regions — is a dangerous practice. It should not to be confused with the distribution of humus-dense organics derived from properly composted excreta.

We, in Western culture, have collectively developed a fear — what Joseph Jenkins refers to as *fecal phobia,* a fear of our own shit. Jenkins's book, *The Humanure Handbook,* (2005) is more than just good bathroom reading; it's a book exploring the philosophy and science of human manure that simultaneously informs, educates, and entertains.

> Our notion of the "smelly outhouse" arises from saturated anaerobic conditions. Waste in this form tends to be unpleasant.

> Excreta = poo + pee + toilet paper

We highly recommended reading it as part of your considerations of CTs. His book dives into culture and science to remove fears and preconceptions around human waste. Research has clearly shown that when the collection, processing, and treatment of these resources is done properly, hazards are reduced and resources are created.

The Questions

Here at home, we have performed hundreds of tours of our systems, and the compost toilet generally piques a lot of interest. People are intrigued, and they wonder about installing one for themselves. However, they have many questions:

- Does it smell? Will there be flies in my house? Does it look gross?
- Can I put toilet paper in the toilet? Do I need to use special toilet paper?
- Can I put other materials in the toilet, like kitchen compost?
- Will rats, bears, or other animals be attracted to my compost pile?
- Do I have to turn my compost pile?
- Should I cover my compost area with a roof?
- What cleaning products are safe for my compost toilet?
- Can I put menstrual supplies, baby wipes, or similar into my compost toilet? (NO! And not in your standard toilet, either).
- Can I build my compost toilet on a second or third floor of my house?
- Is it expensive? How much will it cost? Can I build it myself?
- Are the materials to build a compost toilet easy to find? Where do I find the parts?
- Can I buy pre-built compost toilet kits?
- Can I modify my existing bathroom?
- Do I need electricity?

- Can I have a compost toilet without a fan? If so, how should I design?
- Can I use my compost in the garden for flowers, ornamentals, fruit trees, or vegetables?
- How do I decide which system is right for me?
- Is it legal?
- Can I still use my compost toilet even though I am taking pharmaceuticals? Will cancer treatment drugs harm my compost? Birth control pills?
- How much work is it?
- Can I use wood shavings from my workshop or wood chips from my chain saw?
- What if I don't have straw?

By the end of the book, you will be able to look at the above questions not just as entertainment, but with understanding of the considerations of the system that is most suitable for your needs.

Where to begin? The essential considerations about whether a compost toilet will work for you, and if so, which system to choose, will require evaluating your "needs" and desired outcomes.

Why Choose a Compost Toilet?

The choice to use a compost toilet system over a standard flush toilet (water closet) can be motivated by a variety of factors:

- Water conservation
- Limited or too-expensive access to septic or sewerage services
- Resource recovery
- Financial cost of septic infrastructure
- Absence of electrical supply for on-site septic infrastructure
- Desire for a no-smell bathroom

- Easier-to-clean toilet
- Resilience (less vulnerable to earthquakes, floods, droughts, economic shocks, etc.)
- Philosophical ideology
- Remote location — need to be site-built, using common materials
- Desire to reduce one's ecological impact and carbon footprint
- Wish to discourage visiting relatives
- Wish to educate friends and family

System choice will be guided by the motivations above, and by identifying the purposes and desired outcomes for your compost toilet (e.g. are you prioritizing volume reduction, cost, ease of use, pathogen removal, resource recovery, etc.?).

Some important but typically unconsidered benefits include:

- Proper composting reduces the impacts of pharmaceuticals entering the environment (Carballa et al., 2004; Ternes and Joss, 2008), addressing the massive risk of antibiotic resistance that is magnified by our present water-based conveyance systems.
- Soil containment of the wide distribution of micropollutants (micro-plastics) (Simha and Ganesapillai, 2016).

As you think about your individual reasons why you would like a compost toilet, remember that you can view this with a large lens. Composting of human excreta and returning the sanitized compost to the land may solve a societal time bomb we are just beginning to understand.

Decrease reliance on water

Flush toilets consume 25%–30% of the indoor water consumption of the average home in North America (Canada Mortgage and

Housing Corporation, 2002; DeOreo et al., 2016). Water consumption of toilets poses problems for people not connected to municipal (piped) water, those subject to water restrictions, those who rely on alternate water supplies like rainwater cisterns, or those with low-performing wells. Decreased reliance on water builds resilience, cushioning you from the impacts of water availability fluctuations.

Can a case be made for compost toilets even if you are using piped water? Sure. Piped water has been collected from somewhere, filtered and treated to potable standards, and then distributed through sizeable infrastructure; all stages (collection, storage, treatment, and distribution) require careful management, which comes at a cost. Once flushed, wastewater has to be transported and treated — at a further cost. Both philosophically and economically, there are good reasons to avoid using a precious and highly treated resource to defecate in and flush away.

For those homes that require filtration and treatment of their own water source, more water use means more frequent filter cleaning or replacement. A waterless compost toilet could reduce that recurring cost by 25%.

Recycle nutrients

Conventional sewer disposal systems, which use hydraulics (water flows) to transport waste for disposal, rarely allow nutrient capture, processing, or redistribution to the terrestrial landscape. Large centralized infrastructure, though it has the advantages of efficiency that come with size, is also subject to the toxic waste stream of industry, further complicating the separation of resources. Compost toilets, due to their small scale, allow for a cost-effective and simple method of gathering and processing nutrient-dense resources, with the option of beneficial reuse.

The difference between waste and resource is one of scale. Unused resources become a waste when introduced into the environment at volumes and in practices where ecological systems cannot utilize the nutrients, thus negatively impacting that ecosystem. Those same resources, with a better-managed introduction, can benefit the ecosystem by, for example, aiding in carbon sequestration or feeding the soil.

NOTE: When we discuss reuse of humanure compost, we are not recommending using fully matured and cured compost (defined shortly) on food gardens *unless there is thorough lab testing showing finished compost meets regulated compost standards.* If you do not test, we recommend using your finished product on woody trees and ornamentals; or, it can be buried under 15–30 cm (6–12 in) of cover material. More on this in Chapter 3.

Reduce pressure on septic or sewerage systems

All conventional systems that serve the function of disposing of human waste for single-family residences or small communities rely on infrastructures that have limited lifespans. Reducing septic and sewage flows can in some cases extend the lifespan of an existing system by reducing hydraulic (water) volume on failing distribution fields or treatment plants. Some homeowners may find financial relief in being able to delay or avoid repairs and instead use other disposal systems (including greywater and compost toilets) that minimize the pressures that may otherwise accelerate system failure.

Some jurisdictions charge user fees for the metered amount of sewer water that travels to the regional sewerage system. Cities and towns charge individual homeowners the costs associated with enlarging sewer mains. In some cases, homeowners may find that user fees are reduced or avoided after they install compost toilets.

Site constraints

Some sites have a reduced ability for traditional on-site septic systems, because not all sites have capacity for percolation. Sites that have fractured rock (or little soil) do not offer proper separation of wastewater, which would result in contaminants entering groundwater or surface flows. In instances like this, having a toilet system that avoids requiring percolation is the only option. Properly designed and placed compost piles with absorbent biological mats (of straw, wood chips, peat, or deep soil that allows water to easily percolate through it) can easily handle the moisture from many compost toilet systems.

Emergency resilience

Using a system not dependent on outside sources of water or sewer fits well with resilience/emergency planning. In the event of large power outages or earthquakes, when water and sewer services can be cut off, it's critical to have an emergency backup option to avoid the resulting sanitation disasters that often follow the initial disaster.

Remote locations

Remote off-grid homes, cabins, or lodges are often inaccessible for the installation of a sewer system. On coastlines, it has been common practice to send the sewage pipe into the ocean. In many remote locations, digging a pit toilet might seem to be the only option. But these can be vectors of disease, and they can easily pollute shallow groundwater sources. Compost toilets provide a safe, clean, odorless alternative in all these situations.

Philosophical ideology

Some say that because we don't have flippers or fins, our shit doesn't belong in the oceans. We should take responsibility and not place our waste in someone else's backyard and leave it for some future generation to deal with. Others say, no sense spending more money if you don't have to. Some just want to be off-grid from everything. Some are worried about disasters. Call it the way you see it, but ideology is one of the main drivers in the adoption of compost toilets.

Basic Goals

Whatever your reasons for choosing a compost toilet (CT), and regardless of whether or not you are seeking approval from local authorities, CTs should meet your goals and objectives AND the safety requirements for reducing risks of the following:

- exposure to human or domestic waste
- consumption of contaminated water
- inadequate facilities for personal hygiene
- creating contaminated surfaces
- exposure to contact with vermin and insects

In addition, you need to ensure structural safety for both the user and the structure. All of this and more can be accomplished. Another way of stating this list of requirements would be:

- You don't want gross.
- You don't want soggy.
- You don't want smell.
- You don't want maggots, flies, or rodents.
- You don't want accidents.
- You don't want dangerous.

In short, through good system design and hygienic operational practices we can control for potential pathogen spread to create a safe and pleasant bathroom experience.

The city of Portland Oregon offers a useful publication: "A Sewer Catastrophe Companion: Dry Toilets for Wet Disasters." (Danielsson and Lippincott, n.d., www.portlandoregon.gov/pbem/article/447707)

Certification Standards

The NSF (National Sanitation Foundation, now called NSF International) sets various international standards relating to health and sanitation; it tests products, educates, and provides risk management for the public. For compost toilets, *NSF/ANSI 41: Non-Liquid Waste Systems* is the relevant standard. (ANSI stands for *American National Standards Institute.* It's the organization recognized as the administrative authority for coordinating standards for products and processes.)

NSF/ANSI 41 is a certification for *"composting toilets and similar treatment systems that do not use a liquid saturated media as a primary means of storing or treating human excreta or human excreta mixed with other organic household materials. The standard requires a minimum of six months of performance testing, which includes design loading and stress testing"* (NSF International, "NSF/ANSI 41: Non-Liquid Systems").

The standard does not cover processing, just whether the design can handle the loads and volumes that manufacturers claim. As processing is a minimum two-year time frame, NSF 41 misses the target for determining if the system does what it claims to. This means that a NSF 41-certified product can structurally meet its intended use, but the standard does not assess the functionality of the product to actually process as it may claim to do and thus does not guarantee that materials are decomposed, composted, or safe for disposal. In short, the certification is not intended to determine if a product works.

It costs manufacturers between $15,000–$20,000 per year to test and maintain their NSF 41 certification. This testing, related to volume loading and structural safety, does not cover actual performance or end product. Regulators unaware of the narrow scope of this standard still often use it as a benchmark for accepting manufactured systems and dismissing site-built systems. However, many of the compost toilet manufacturers have dropped their NSF certification, instead relying on evolving performance-based guidelines (such as those introduced in the Province of British Columbia) that show how to meet all the objectives and functions as required in the codes (Lippincott, 2010).

The takeaway point is that NSF 41 is not a standard proving function and performance. If you ever need to verify a manufacturer's claim, you can use the calculations given in Chapter 4.

There is another certification that is often shown on the label of composting toilets. It is the ETL 3086410, which means the product conforms to UL 1431 (Standard for Personal Hygiene and Health Care Appliances) and CSA C22.2 No.68-92 (Motor-Operated Appliances for Household and Commercial). Retailers are promoting confusion as to this being the appropriate standard to apply to a particular style of toilet. In fact, this standard is tied to the electrical safety of a broad range of appliances, from shavers, massagers, garbage disposals, and other small electrical appliances. Thus, the ETL certification does not specify the functionality of the toilet, but is directly tied to its electrical safety.

Again, the functionality of a toilet is best looked at objectively, and designed to meet the needs of a specific application.

One final certification to mention is the very recent IAPMO WE•Stand (Water Efficiency and Sanitation Standard). IAPMO is the International Association of Plumbing and Mechanical Officials, a group that provides code development assistance, resulting in such items as the Uniform Plumbing Code and Uniform Mechanical Code. The WE•Stand is a performance-based standard released in November 2017 as an American National Standard covering water efficiency for both residential and non-residential applications. Within this standard are provisions for composting toilets (among many other items, including greywater and rainwater), with the first set of comprehensive codified requirements for the installation, safe use, and maintenance of composting and urine diversion toilet fixtures. Requirements include separate collection devices (commodes), and compost processors that are covered and vermin proof, durable construction materials, proper handling of fluid (leachate) discharges, and discharge requirements. The first system to apply this standard was a humanure system in Portland, Oregon, in May 2018. This standard closely mirrors the Provinces of BC's guidelines in the Manual of Composting Toilets and Greywater Practice released in September 2016.

Objectives

Do your objectives and goals prevent you from meeting the objectives and goals of regulators? As noted above, homeowners may have their own rationale for choosing a compost toilet, such as personal philosophy or costs. When owners attempt to initiate actions to meet their goals, they often bump into inspectors with an extremely narrow and prescriptive understanding of building codes and health regulations. Often, there are poor outcomes, and even conflict.

Obviously, if you are the homeowner or the contractor, you want to ensure the safety of everyone using a system, and you want to avoid polluting groundwater or gardens. You also wish to avoid odors, fire risks, and poor hygiene. It is important to understand that the regulators' mandate is to focus on a narrow subset of your objectives; their mandate is not the big picture. The big-picture items are outside the jurisdiction of a building official or health officer, and their training is around the details within their focused scope.

The essential takeaway point is that it is up to those wanting to design and use a compost toilet to understand the role of the regulator and to clearly identify how your choices to incorporate such a system meets the objectives of the regulators. It comes down to understanding and good communication.

Figure 1.1 shows the different perspectives of homeowners and regulators. The homeowner's bigger picture is inclusive of the narrower focus of the regulators'.

If you choose to have your system permitted, how do you communicate these broader goals to officials? To begin with, it is perhaps wise to stop and look at the history of building codes and how they have evolved. Most codes are created to address

Fig. 1.1: Regulatory Code Lens. Risks as viewed by the regulators is a subset of a wider set of risks as they might be viewed by the homeowner, designer, or contractor.

Illustration credit: David Eisenberg, DCAT, 2010

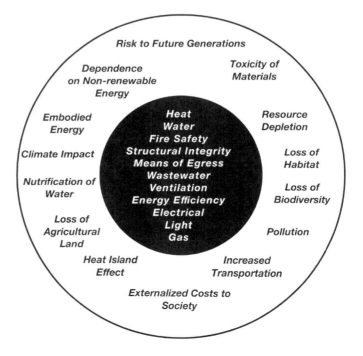

accidents or injury that already occurred; because they are based on past failures, they instruct us what NOT to do. Early codes tend to be prescriptive in nature (specific and precise), in essence providing lists of check-boxes for the inspector. If a box couldn't be checked off, the item in question would not be allowed. Europeans began to realize that this approach stymied innovation and creativity and thus began moving away from prescriptive codes to performance-based ones, wherein you had to demonstrate how something performed. This helped Europe become a leader in building innovation, witnessed in both the success of Passive Haus and the implementation of modern compost toilets systems becoming commonplace. In North America, there was also this understanding that the codes needed to be modernized, but there was a stronger attachment to the prescriptive nature, therefore regulators (particularly those in Canada) began to investigate how to mesh the prescriptive and performance-based approaches. The outcome was *objective-based codes* (Potworowski, 2010).

The intent of objective-based codes was to create regulations that allowed innovation and creativity while still giving a minimum standard that had to be adhered to. These minimum standards were originally the documented rules in the prescriptive code but are now called "acceptable solutions." What a concept! By listing minimum standards (acceptable solutions) and by stating the objectives and the functions that had to be met, the door opened to *alternative solutions,* — solutions that meet the intent of objectives, rather than the letter of the law.

Today's codes set out the WHY (objectives) and WHAT (functions) and provide us the opportunity to present the HOW (solutions).

In summary, there is now opportunity for deviation from older codes to new, novel, and innovative "alternative solutions." When we can demonstrate that the rationale of an alternative solution meets the objectives and intents of code, then we can be allowed to implement new ideas while still providing a reasonable degree of risk control.

Learning the language a regulator might use may facilitate better communication and head off problems before they begin. In light of this, the rest of this section gives a basic overview of objectives and functions as understood by a regulator.

Objectives lists

Objectives are the WHY. Below are some of the objectives your inspector will be considering when evaluating your project — in the language they use in their codes.

- **Safety** — Design should limit the probability of exposing any person in or around a building to unacceptable risk or injury.
  - ○ Structural Safety — Design should ensure that the system can bear the weight loads placed on it, and that the system does not place undue loads on other aspects of the structure, which could cause collapse or failure, or cause deterioration of the building elements.
  - ○ Fire Safety — Design should ensure that the system does not create a fire or explosion risk, or impair the functioning of a fire suppression system, or impede access of evacuation.
  - ○ Safety in Use — Ensure that the system limits the probability of slips, trips, falls, contact with hot surfaces, and exposure to hazardous substances, and that hazardous substances are fully contained.

- **Health** — Design and construction should limit a person's exposure to unacceptable risk.
  - ○ Healthy Indoor Environment — Ensure the design of the system ensures good indoor air quality, a lack of contact with moisture, and adequate thermal conditions.
  - ○ Sanitation — To ensure that the design and install of the system does not expose those who use it to human waste, cause or create the opportunity for the consumption of contaminated water, contact with contaminated surfaces, lack of access for personal hygiene, or contact with vermin or insects.
  - ○ Noise and Vibration Protection — Ensure that people are not exposed to dangerous levels of noise or vibration.
  - ○ Hazardous Substances — Ensure that people in or around a building are not exposed to hazardous substances.
- **Protection against Water and Sewer Damage** — To ensure the system does not leak water or sewer/septage.
- **Energy Efficiency** — To ensure that the system does not negatively impair the energy efficiency of the building.
- **Water Efficiency** — To ensure that the system demonstrates higher water efficiency than the standard water toilet/closet system.

Functions

Functions are the WHAT. It's the thing that delivers functionality as required under the various objective-based codes (Building, Plumbing, and Fire). Our alternative toilet solutions need to explain HOW we will address WHAT the code requires. The following is only an example of a few *functions,* as one would see in a building code. The way

to read the subsequent phrases is to follow up each statement with the action.

Example: To resist the entry of vermin and insects...

I would screen all vent pipes, reduces excess moisture by actively venting with a fan and diverting urine, and seal off compartment doors with snug weather stripping.

- To minimize slipping, tripping, falling, contact, drowning, or collision…
- To minimize contact with hot surfaces…
- To limit the level of contaminants or the generation of contaminants…
- To minimize the risk of release of hazardous substances or spread beyond their point of release…
- To minimize the risk of contamination of potable water…
- To limit moisture condensation…
- To provide facilities for personal hygiene…
- To provide facilities for the sanitary disposal of human and domestic wastes…
- To minimize the risk of malfunction, damage, tampering, or misuse…
- To minimize the risk of inadequate performance due to proper maintenance or lack thereof…

You need to find out what ordinances and codes are relevant in your locale. In the U.S., one tool is the MuniCode, https://library.municode.com; follow the links from State to City/County. In Canada, no such tool exists; you will need to directly contact your local government.

It's like when traveling in a foreign country: if you make the effort to learn the basic aspects of the language — and others see you working at it — you are likely to have a better outcome in your communications.

Collection Systems: A Brief Overview

This section is a brief introduction to the different types of systems. They are discussed in more detail in Chapters 5 through 7. Although this results in some duplication, many readers will appreciate this early introduction.

There are many manufactured brand-named toilets available to you (including discontinued brands). In our review, we classified toilets based on how they function, and we picked just a few brand names to use as examples. There are many manufactured compost toilets on the market, so it would not have been practical to include all their

brand names. The goal is to understand the basic categories and then seek out locally available products and systems to meet your needs. Many countries and geographic locations (i.e. North America, Europe, and Asia) will have different names for exactly the same system or even the same manufacturer.

Compost toilets systems are separated into two categories: *batch* and *continuous*. All systems follow the same general principles of material flows as seen in Figure 1.2.

Batch Systems

Batch systems generally collect raw materials into a receptacle (a bin or bucket), which, when full, is removed from the collection

Fig. 1.2: *Material Flow Pathways. There are a host of directions materials can flow, and choice of system will determine which pathways are followed.*

ILLUSTRATION CREDIT: GORD BAIRD

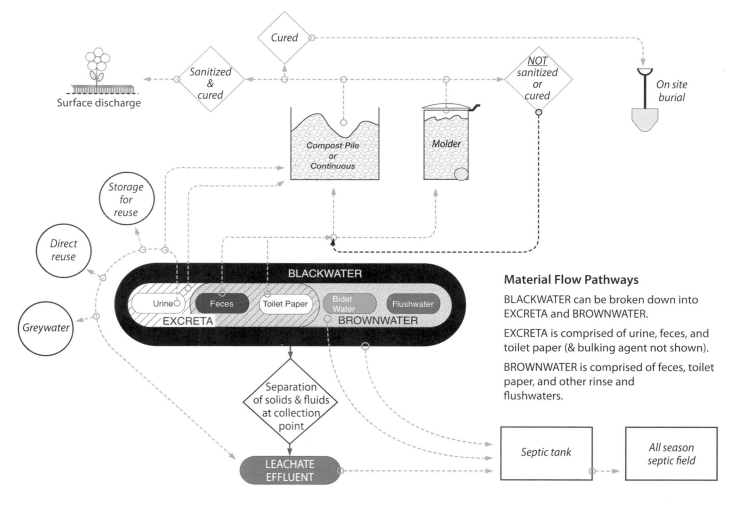

Material Flow Pathways

BLACKWATER can be broken down into EXCRETA and BROWNWATER.

EXCRETA is comprised of urine, feces, and toilet paper (& bulking agent not shown).

BROWNWATER is comprised of feces, toilet paper, and other rinse and flushwaters.

area. The contents of the full receptacle are then processed. The processing for batch systems can either involve emptying the bin into a compost pile for immediate composting (commode batch), or one can store the contents in the very same bin for long periods of time (chambered batch, or moldering systems). Either way, these batch systems all have a distinct separation between the collection and the aging processes.

Commode Batch Systems

Commode batch systems, commonly referred to as the *humanure* or *bucket system,* have smallish 20 L (5 gallon) buckets under the pedestal (the commode). When a person leaves a deposit, they "flush" by adding sawdust to cover the deposit. The buckets fill up regularly (constituting a *batch*) and are swapped out routinely; full ones are stored until there are enough to make dumping them into the compost pile worthwhile (4–12 buckets). Some commode batch systems use larger, wheeled bins — up to 80 L (20 gallons).

The active compost pile is "added to" for a year; then it is left to sit dormant for a year, allowing it to cycle through the composting stages.

Commode batch systems are discussed in greater detail in Chapter 5.

Chambered/Moldering Batch Systems

In chambered/moldering systems, material is left to sit for extended periods of time (thus, it *molders*). Chamber/moldering batch toilets are different from the commode batch in that the collection receptacles are much larger (bigger *batches*), and they usually involve urine diversion and leachate drainage. They are not transferred and dumped into a compost pile once they are filled. Instead,

the solids are held in the collection chamber or bin to age in place for one or more years. Only after this time has passed are the contents either placed into a compost pile (if more processing is needed) or used as a soil conditioner. Moldering systems tend to have a couple of basic configurations: they either incorporate large containers (200 L/45 gal or larger); or they are stationary chambers/vaults that are integrated into a building's design.

Chambered/moldering batch systems are discussed in greater detail in Chapter 6.

Continuous Systems

In continuous systems, new materials are collected and a degree of decomposition is undergone, all within one unit. There are two basic types:

- **Self-Contained Continuous** — small, all-in-one units designed primarily for cabins, seasonal, or recreational use.
- **Centralized Continuous** — large systems designed for regular daily usage.

Despite there being a degree of decomposition within the unit (to the level a manufacturer might call "finished,"), the materials are *not* matured or sanitized (maturity/sanitation is discussed in Chapter 2). Any system that receives continuous raw inputs offers the opportunity for nitrogen and pathogen-rich liquids to re-contaminate all materials in the chamber regardless of their state of decomposition. For this reason, we will continually restate the requirement that "finished" materials from a continuous system receive batch processing (composting) to make them safe.

Many systems incorporate mechanical mixing devices of one form or another (i.e. rotating drums, tines, scraper arms) to aerate and move materials from the raw stage to a stabilized (partially matured) stage and into a location or compartment that can be accessed for removal. These systems can be more complicated because they incorporate active components to speed processing (pumps, heating elements, motors, etc.).

Moisture management for these systems can include any combination of the following:

• urine diversion
• heating elements
• fans and ventilation
• leachate drains

Sizing these systems to match usage patterns is critical because these systems can only decompose materials and eliminate moisture at a specific rate (which is often temperature dependent). If new materials are added beyond its processing capacity, the system will fail. Manufactured systems come with recommendations for the number of continuous daily users the system was designed for.

Continuous systems are discussed in greater detail in Chapter 7.

Vermicomposting Systems

Vermicomposting has aspects that allow its incorporation into both continuous systems and batch moldering systems. Worm introduction requires good moisture control and can offer more complete decomposition within shorter time frames. The speed of decomposition does not greatly impact sanitation, but it enables a system greater flexibility in receiving higher use (more people) and can reduce the need for servicing a system.

After materials are processed by worms, they are highly stabilized; the materials can be buried or transferred to a compost processing pile for further maturing and pathogen attenuation. "Highly stabilized" refers to the creation of a material that is very homogeneous in appearance and has low nitrogen and other phytotoxins, making the materials safe for plants.

Vermicomposting systems are discussed in greater detail at the end of Chapter 7.

Chapter 2

Safe Composting

THE MOST CRITICAL ASPECT OF COM-POST TOILETS is the composting process. There are many important concerns. Will it leach pathogens into the environment or my well? Will it attract vermin? Will it smell? Will dogs get into it? Will kids play in it? Is it safe? Will the local authorities take my kids away?

The answer to all these could be "yes" if the composting system is not designed well. What it comes down to is controlling for *moisture, oxygen, time,* and *temperature,* and, perhaps most importantly, *controlling for pathogen containment.*

Compost Life

Composting is the process of breaking down organic materials through processes of decay and digestion within an oxygen-rich (aerobic) environment. The primary organisms at work in the decomposition and composting processes that degrade pathogens includes *microorganisms* and *soil animals.* Microorganisms include bacteria, actinomycetes, fungi, algae, slime molds, yeasts, viruses, and lichens. Soil *animals,* which aid in aeration, bacterial predation, and degrading surface litter, include protozoa, earthworms, arthropods, amoebas, and nematodes. How fast these organisms work is dependent on available nutrients, environmental factors, and the number and health of the organisms.

These organisms are the workhorses that break down the original materials to a variety of qualitative states (*raw, stable,* and *matured,* as discussed below). Microbes play a larger initial role in processing from the raw to stabilized states, while soil animals play more of a role in the maturing process. Conveniently, the conditions that support and enhance composting organisms' health are generally the same conditions that promote and enhance pathogen death. If you enable a healthy composting process, sanitization will naturally occur over time.

Inconveniently, the chemicals we use on our bodies and for cleaning are often *anti-life.* We need to avoid using anti-bacterial soaps, ammonia, chlorine, and other harsh products in and around ourselves and our toilets if we want compost organisms to flourish.

Critical Environmental Factors

Aeration — Without oxygen, the organisms we want to flourish will die, so it is imperative to create conditions for good air exchange in each stage of the process. What you are trying to do is create mechanical means of entry for oxygen and escape for the carbon dioxide given off by decomposition. Achieving a good surface area-to-volume ratio through a combination of a loosely textured material and air pockets (pores) within the pile is important. A good ratio can be created by the regular addition of coarse bulking agents (like sawdust) or the introduction of earthworms in the advanced stages. It is also possible to use mechanical means to stir the pile to add oxygen. Increasing air flow by actively venting the material aids in gas exchange (and evaporation of excess moisture). If the materials become too wet, then the open pore spaces don't exist, and anaerobic

Pathogens can be bacteria, worms, amoebas, protozoa, viruses, fungi, or prions. All are potentially harmful to humans.

conditions occur — leading to a *different* population of bacteria (*anaerobes*), which release noxious gases like hydrogen sulfide, ammonia, and methane gas. Yuck!

Moisture — A moisture content between 45%–70% (a wide margin) is the target. Below this range, it is too dry for organisms to thrive; above this range, the pores begin to fill with fluid, creating those anaerobic (without oxygen) conditions just mentioned.

Temperature — Different temperatures support different biological lifeforms, activity, and processes (e.g. earthworms won't survive high temperatures of 45°C to 80°C [113°F to 176°F]). The Q10 temperature coefficient applies to composting: for every 10°C (18°F) in temperature rise, the biochemical activity doubles — until pasteurization temperatures are reached, at which point biological activity drops off. Pasteurization is not defined by a particular temperature per se, but rather as a temperature range that results in the death of organisms exposed to it over a certain period of time.

In ideal compost conditions, the compost pile hits both the running-out-of-food limit (carbon) and the building blocks limit (nitrogen) at the same time. Achieving the right balance is surprisingly easy: the carbon is derived from the poo, toilet paper, and bulking additives, and the nitrogen comes from our urine. When limits are hit, temperatures drop, which allows the next stages of the process to occur: soil animals and fungi flourish and carry out the long-term aeration and the conversion of lignin (a complex carbohydrate found in plant cell walls), and the remaining carbon into simpler forms (Del Porto and Steinfeld, 1999).

There are three main classifications for microbial communities; they are grouped together by the temperature ranges in which they survive and thrive. For the purposes of composting/decomposition, each group acts differently and serves a different purpose.

Psychrophilic: -10°C to 20°C (14°F to 68°F)

- In low temperature decomposition, *psychrophiles* dominate. Predominant organisms include actinomycetes, fungi, and larger soil animals (like worms and arthropods).
- Pathogen attenuation (reduction) is greatly inhibited. In some cases, pathogen *preservation* results — in the same way we use a fridge to preserve food (the cold doesn't kill bad things, it just slows them down).
- There are few cool-temperature psychrophiles that are human pathogens.
- Cool-temperature conditions can be found in "moldering" systems and in late-stage curing of matured composts.
- Composts that never exceed this temperature profile are not considered sanitized; they therefore require testing before direct burial at 30 cm (1 foot) or deeper. Otherwise, they require thermophilic (high-temperature) treatment or other sanitization process to be considered safe.

Aerobic compost smells sweet; anaerobic smells foul.

Above 75% moisture, you would be able to squeeze material and have it drip — not that we are suggesting you do this of course, but this is a method of testing applied to conventional composts.

The warmer the temperatures, the bigger the bacterial compost party — until they cook themselves (kind of like human-induced climate change). In composting toilets, given the right environmental factors, early processing will self-generate temperatures that promote more biological activity; that activity creates warmth. At some point, one or more of the limits to growth are hit. These limits include a too-high temperature, running out of food, a shortage of water, or lack of oxygen. Available nitrogen also plays a critical role in that it allows bacteria to build cell walls and to multiply; if nitrogen levels drop, bacteria run out of the building materials they need to reproduce.

Mesophilic: 20°C to 44°C (68°F to 112°F)

- In warm temperature decomposition, *mesophiles* dominate. Predominant organisms include bacteria, fungi, actinomycetes, and some soil animals like protozoans, rotifers, and worms.
- Mesophilic activity is supported by warm ambient air temperatures or external heat sources.
- Mesophilic conditions are the most common in all types of compost toilets and processors, they precede and follow the thermophilic temperature phase.
- Composts that do not exceed this temperature are only considered sanitized after a minimum of two years; at that point, they can be directly buried at 30 cm (1 foot). Otherwise, they require further thermophilic treatment or other sanitization process to be considered safe.

Thermophilic: 45°C to 80°C (113°F to 176°F)

- In this high temperature range, *thermophiles* dominate.
- Most thermophiles are bacteria; soil animals do not survive such high temperatures.

- The regulatory agencies that set composting standards and guidelines define true composting as one that undergoes biological processes at these high temperatures.
- Commonly called *hot composting*. Pathogen death rate increases as temperatures rise.
- Ideal conditions for hot composting require a carbon-to-nitrogen balance (C:N ratio) that is roughly 30:1 (carbon for food, and nitrogen for amino acid and protein production). Too much carbon slows the process down and causes a cooler pile; too much nitrogen will cause unpleasant odors (Richard and Trautmann, n.d.).

An adequate volume of material ($>1 \text{ m}^3$ or $>1 \text{ yd}^3$) is required to retain the self-generated heat from the bacterial activity, retain moisture, and insulate the contents from the cooling influence of ambient air temperatures (this volume will protect compost temperature even in freezing temperatures). Because such conditions are rare in nature, thermophilic compost doesn't naturally exist (except in relatively rare situations). Additionally, diversity supports more microbial activity, more oxidization, and more heat.

If you have a smaller compost pile, you can facilitate decomposition by adding straw

Fig. 2.1: *Thermophilic temperatures signal a very biologically active compost pile, creating a hostile environment for pathogens and a rapidly stabilizing mass.*
PHOTO CREDIT: GORD BAIRD

We often start a new pile with a dead rodent, chicken, or something with intact intestines, then we add all our kitchen scraps; this adds to the initial microbial life present, allowing the pile to more rapidly reach hot temperatures.

Fig. 2.2: *Comparison of compost temperature profiles.*

ILLUSTRATION CREDIT: GORD BAIRD

insulation and/or more garden materials, putting a seasonal roof over the compost — or having more kids to help make more compost.

In summary, thermophilic conditions are the key defining aspect of what delineates true composting from decomposition. Mesophilic bacteria thrive in the most common temperature range on the planet, including in the body temperature range of most mammals. Psychrophilic bacteria like it cool and function at temperatures not associated with pathogen death; in fact, they aid in preserving them in a dormant form. Compost *toilets* have psychrophilic and mesophilic phases; compost *piles* will transition through all three temperature ranges.

Temperature Profiles

If we were to chart the temperature of a pile of raw materials from the start of new additions till the end, when everything

is decomposed, we would witness a pattern. The pile starts out cool, but quickly heats up as biological life becomes active. Temperature peaks when resources start to decline (like nitrogen, carbon, water, and oxygen), then there is a long cooling stage (see Figure 2.2).

Snapshot of the ideal compost conditions:

- Carbon-to-nitrogen ratio of 20–35:1
- Moisture content: between 40%–60%, and not above 75%
- A pH between 5.0 and 8.0 (pH 0 [acid], pH 7 [neutral], pH 14 [basic])
- Porosity, or air spaces: 30% of the materials should have their surface area exposed to air to allow for air flow and gas exchange; particle sizes should not be too small because that allows compaction. You want a fluffy compost.

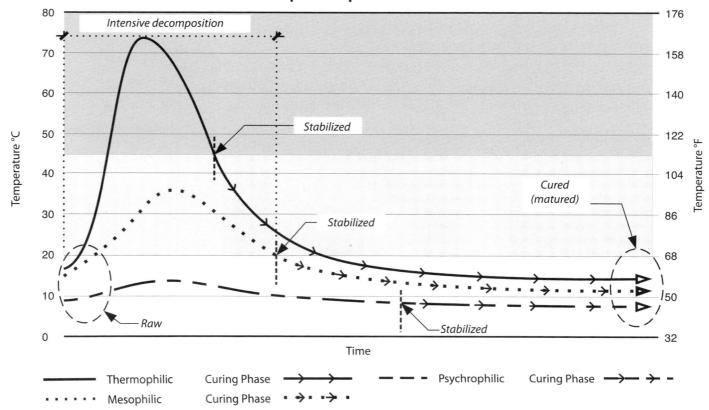

Materials and Processing States

There are three distinctive processing states that can be differentiated by their physical qualities. These are *raw, stabilized,* and *matured,* and they are linear in progression. The compost toilets system employed will determine the progression as various toilets manage the critical factors just discussed differently; they rely on different temperature profiles to carry out their processes.

Raw Materials: Excreta

As an adult reading this book, your mental image of what raw materials are and look like is likely pretty accurate, so we will spare you the details. Well, maybe not. For our purposes, we often refer to raw materials as *excreta,* which includes feces, urine, and toilet paper that has not undergone any form of processing related to time and/or temperature. This raw stage is where there is the highest opportunity for pathogen transfer and potential for nitrogen to leach from soils into waterways (Wichuk and McCartney, 2010).

Raw materials also include other additives. Usually, you will add to the excreta some type of biodegradable bulking agent, like peat moss, chipped leaf mulch, fine wood shavings, coarse wood sawdust, or shredded toilet paper tubes.

In the thermophilic compost, there are other raw materials that are often included (some manufacturers specify it) — items that are not normally appropriate for your garden compost, such as bones, meat scraps, or egg shells. Surprisingly, in thermophilic conditions, such items are consumed rapidly by bacteria; we, ourselves, have witnessed dead chickens completely consumed within four days, and animal processing yards use a thermophilic system to address their wastes. However, for those compost toilet processing systems that rely on moldering, these items will not be appropriate.

Looking deeper (sorry, this is important), feces is roughly 75% moisture and 25% solids. Figure 2.3 illustrates the breakdown even further. The solids are two types of solids: the volatile organic solids (carbon-based solids that, when dried, will burn, including proteins, fats, carbohydrates, dietary fiber, and total nitrogen); and fixed solids, also known as minerals, which do not decay.

Urine is 96% moisture and 4% solids, wherein the solids are primarily fixed (mineral); Figure 2.4 shows the breakdown of what's in your urine.

Fig. 2.3: *What's in your feces? Composition of a turd.*

Illustration credit: Gord Baird

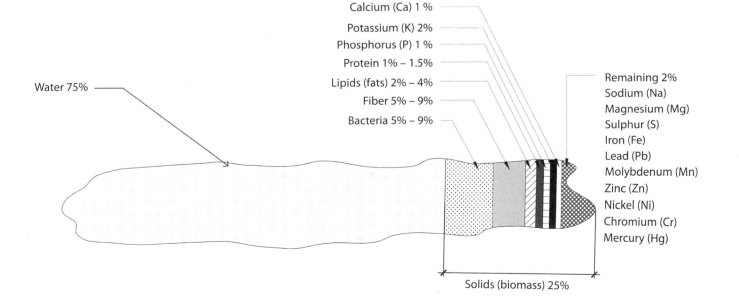

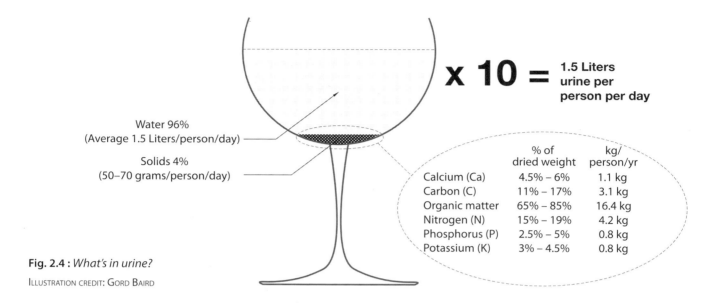

Water 96%
(Average 1.5 Liters/person/day)

Solids 4%
(50–70 grams/person/day)

x 10 = 1.5 Liters urine per person per day

	% of dried weight	kg/ person/yr
Calcium (Ca)	4.5% – 6%	1.1 kg
Carbon (C)	11% – 17%	3.1 kg
Organic matter	65% – 85%	16.4 kg
Nitrogen (N)	15% – 19%	4.2 kg
Phosphorus (P)	2.5% – 5%	0.8 kg
Potassium (K)	3% – 4.5%	0.8 kg

Fig. 2.4 : *What's in urine?*

Illustration credit: Gord Baird

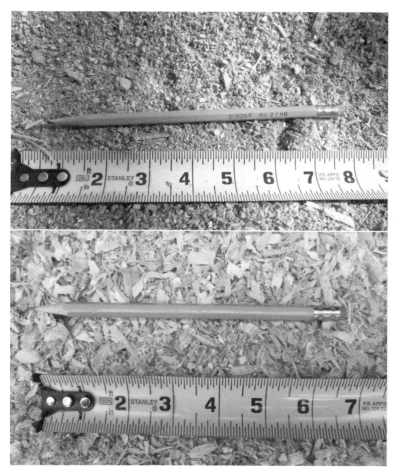

Fig. 2.5: *Raw Materials: Coarse sawdust/shavings.* Photo credit: Gord Baird

- Ninety percent of the nitrogen that is excreted from the body is in the urine.
- 1.5 L (1.6 quarts) of urine is excreted on average per person per day (10 full wine glasses).
- 0.18 L (0.19 quarts) of feces is excreted on average per person per day.
- More solids come out of your urine (4% of 1.5 L) per day than your poo (25% of the 0.18 L).
- Of all the solids or biomass, 25%–50% is bacterial bodies. SERIOUSLY!
- Feces is pH 6.6 (neutral) and urine is pH 6.2 (slightly acidic).

The photos in Figures 2.5 through 2.10 provide visuals of ... stuff. They clearly show what different materials look like and how they are deposited into the center of the compost pile. Figure 2.10 shows the appearance of the pile with the cover materials added and thermometer installed. The compost pile in these pictures ranges between 60°C–76°C (140°F–170°F) at peak temperatures throughout all four seasons.

Fig. 2.6: *Raw Materials: Kitchen compost including vegetables, egg shells, coffee grounds, and meat and bones.* PHOTO CREDIT: GORD BAIRD

Fig. 2.8: *Raw Materials: Batch commode bucket two-thirds full — visually more appealing than the kitchen compost and less fragrant.*

PHOTO CREDIT: GORD BAIRD

Fig. 2.7: *Raw Materials. Kitchen compost is the first component to be dumped into the pile — unsightly and fragrant.* PHOTO CREDIT: GORD BAIRD

Fig. 2.9: *Raw Materials: Collected excreta added atop kitchen compost — visibly high in carbon (sawdust/shavings).* PHOTO CREDIT: GORD BAIRD

Fig. 2.10: *Raw Materials: The new additions are flattened with a dedicated rake and covered with 6"–8" of loose straw (or spent animal bedding, leaves, shredded paper, etc.).* PHOTO CREDIT: GORD BAIRD

Raw = unprocessed excreta

Stabilized = materials that have low-to-no odor and unrecognizable components

Mature = stable forms of nitrate with no ammonia, decreased and stable CO_2, and no odor

Sanitized = safe for humans.

Raw materials for large batch chamber compost toilet systems (aka moldering systems) or continuous system are visibly very different, as pictured in Figure 2.11. These systems employ moisture diversion, therefore less carbon bulking agents are required.

Fig. 2.11: *Raw Materials: Raw excreta deposited into a Terra Nova compost toilet, viewed through the rear inspection port.*
PHOTO CREDIT: A. SCHÖPE, SUSTAINABLE SANITATION ALLIANCE

Fig. 2.12: *Stabilized Materials: No identifiable constituents of the raw materials are visible — there is no odor, but the material is still potentially a carrier of pathogens.*
PHOTO CREDIT: GORD BAIRD

Less bulking agents make the pile less aesthetically pleasing, but in these systems you don't have to look at it, so who really cares? Even though bulking agents are not required to absorb moisture in these large systems, they are often still applied to provide porosity (air spaces) to the materials.

Compost Stages
Stabilization

Stabilized Compost Compost toilets start with the collection of raw, unprocessed materials. These organic materials have high levels of nitrogen, bacteria, and potentially dangerous pathogens. As these materials decompose, they undergo biochemical changes that break organics into smaller, more homogeneous pieces, reducing the pathogens and converting the nitrogen to a state with decreased phytotoxicity (toxicity to plants). Eventually the materials reach a stabilized state, qualitatively characterized by having no discernable original materials, low-to-no odors, and a high nutrient content to support moderate biological activity. This stage will likely still have pathogens. Only the highly stabilized versions will not tie up nutrients in the soil or reduce oxygen availability (Ralston, 2016). Stabilized materials are NOT yet ready for dispersal.

Stabilization can be achieved through any of the three processing temperatures, but it usually occurs in the mesophilic and thermophilic ranges. However, a longer time at lower temperatures will also achieve stabilization.

Stabilization is also marked by the quality of resisting further decomposition. As the microbial activity consumes the carbohydrates and nitrogen, activity slows, resulting in less decomposition (Wichuk and McCartney, 2010).

Maturation

Mature (Cured) Compost Microorganisms continue to convert the organics to smaller particles, as these organics are their "food source." In the presence of this abundant food, the microorganisms grow and reproduce, using the volatile forms of nitrogen (urea and ammonium) to build their cell walls, and in this process are converting these volatile forms of nitrogen into stable, safe forms for plants (discussed in Chapter 7). Microorganism activity slows when resources decline, due to decreased food (carbon) and nitrogen, which is indicative of completed decomposition. At this stage, the contents rest at ambient temperatures (psychrophilic and mesophilic), in an aerobic condition, for a specified time.

Mature compost is qualitatively characterized by homogenous small particles and no odor; nutrient sources have been consumed to a point where biological activity has slowed and stabilized. This can be demonstrated through testing (discussed in depth later). The mature stage ensures that materials are safe for plants and can be introduced into the soil, though it still does not mean that all potential pathogens are dead (i.e. the compost is not *sanitized*) (Ralston, 2016; Wichuk and McCartney, 2010).

The maturation pathway is slightly different for thermophilic and non-thermophilic processes (see Figure 2.13). In thermophilic processing, maturation starts when the temperatures drop down into the mesophilic range, and it continues over time as long as pile temperatures are ≥ 5°C (41°F). In non-thermophilic piles, maturation is defined by the period of time at a particular temperature. During this long period, the cooler temperatures allow for insects and fungi to enter the pile and continue biological processing. Insects mechanically break down materials — they literally chew it all up. But more importantly, they re-create tunnels and airways throughout the compost pile to allow for gas transfer (primarily oxygen

Fig. 2.13:

Maturation pathways for thermophilic and non-thermophilic processes to be considered cured or matured and ready for burial only.

ILLUSTRATION CREDIT: GORD BAIRD

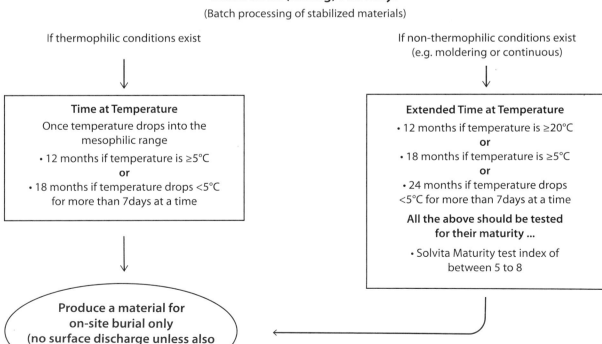

Maturation (Curing) Pathways
(Batch processing of stabilized materials)

If thermophilic conditions exist

If non-thermophilic conditions exist
(e.g. moldering or continuous)

Time at Temperature
Once temperature drops into the mesophilic range
• 12 months if temperature is ≥5°C
or
• 18 months if temperature drops <5°C for more than 7days at a time

Extended Time at Temperature
• 12 months if temperature is ≥20°C
or
• 18 months if temperature is ≥5°C
or
• 24 months if temperature drops <5°C for more than 7days at a time
All the above should be tested for their maturity ...
• Solvita Maturity test index of between 5 to 8

Produce a material for on-site burial only (no surface discharge unless also confirmed sanitized)

and carbon dioxide). Fungi also work on the woody cellulose to break down the lignin, thus playing a huge role in further homogenizing the materials, as seen in Figure 2.14.

Again, it is important to reiterate that despite certain qualities defining a compost as mature, or cured, the compost is still not sanitized. Sanitization processes are extremely likely to coincide with maturation processes, but we cannot always assume this.

Sanitization

The final procedure, sanitization, is dependent on the time frame the materials are held at given temperatures. This process can be further manipulated by some other action such as the addition of ammonia, changing the pH, or pasteurizing with external heat or other chemical process. Ultimately, *sanitized* means that the pathogen levels are below the threshold to cause disease. Just as *cured* does not ensure sanitization, sanitization does not ensure maturation (you could conceivably take raw materials and heat them up to kill pathogens, thus sanitizing it, but you could still be leaving high levels of nitrogen that would be unsafe for plants).

All types of pathogens (viruses, bacteria, fungus, protozoa, and parasitic worms, or helminths) (Alberts et al., 2002) will die through one of the several pathways. The process of reducing a pathogen load is called *pathogen attenuation,* which in regular language means *killing the disease-causing bugs.* When pathogens loads have been reduced to levels that are no longer dangerous to humans (making the compost safe for handling and discharge into the environment), we call this *sanitized.* But sanitization is NOT to be interpreted as disinfection or sterilization; sterilization is the elimination of virtually all microbial life.

To understand the sanitation process, we need to briefly touch on pathogens and how they spread.

Pathogen Spread

There are many routes by which pathogens spread. Our primary goal to stop any possible contact of feces with the face; to do this, we use *the multi-barrier approach.* (See Figure 2.15 showing the "F Diagram.") The multi-barrier approach is used to:

- Avoid direct contact with human excreta.
- Limit vector-borne transmission via rodents, insects, birds, pets, or mischievous small children.

Fig. 2.14: *Matured Materials: More finely textured than the stabilized materials and no trace of phytotoxic compounds.*
Photo credit: Gord Baird

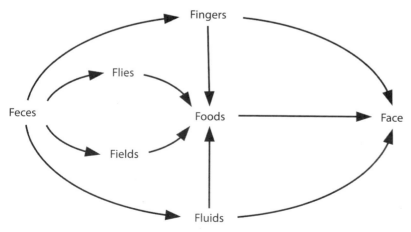

Fig. 2.15: *The "F Diagram" summarizes the main ways fecal pathogens are spread from the feces to the face.* Illustration credit: Gord Baird, adapted from Winbald et al., 2004

- Ensure no inhalation of dusts or aerosols from use of materials before fully treated.
- Avoid misapplication of unsanitized materials on food crops.
- Avoid leaching pathogen-rich fluids into waterways that would affect drinking water or foods harvested from those waters.
- Stop domestic and agricultural animals from consuming human excreta and posing as new hosts or transmission sources.
- Keep facilities clean.
- Promote personal hygiene, halting secondary spread among the population.

By *barrier* we don't just mean physical obstruction. A barrier can also be an action or an approach. All of the following can be part of a multi-barrier approach: personal sanitation, toilet design, vermin control, leachate management, compost confinement, temperature/time and moisture management, and safe-handling practices. You may notice that these barriers are reminiscent of the regulatory objectives in Chapter 1.

Here are some barriers to employ that prevent pathogen spread:

- **Don't allow animals to gain access to unsanitized materials** (e.g. dogs, pigs, cattle, or birds). Using seagulls as an example, we know that tapeworm can be transmitted from human feces to birds, whose droppings can land on roofs and enter rainwater cisterns. If the birds cannot gain access to unsanitized human feces, then you have a sanitary barrier. Keep creatures out of your compost pile.

- **Don't allow human feces to enter the soils or water systems unless treated.** Some worms require fish or snails as a host. Remove the possibility of aquatic contamination by collecting and treating leachate or by using absorbent biological sponges under compost piles. If excess moisture will be an issue, roofing over the compost bins will avert a waterlogged sponge.

- **Insect vectors can be easily reduced.** Inside the area that houses the collection receptacle, the use of specialized flytraps can help limit their populations. (See Figures 6.15, 6.16, and 7.6.) Always ensure that there is no pooling liquid by using proper drainage; keep buckets and bins tightly closed until composting time; keep lots of cover materials on compost piles, and consider covering the entire pile with some sort of roof structure; use appropriate safety mask and gloves when dealing directly with the composts.

- **Heat is another form of sanitary barrier.** Flies are attracted to fresh feces, and they lay their eggs in it; however, adults, larvae, and eggs cannot survive temperatures above 122°F (50°C) — another good reason to use thermophilic composting.

Pathogen death

Pathogens are classified based on three qualities:

1. Virulence — severity or harmfulness of a pathogen
2. Latency — period of dormancy of a pathogen before acute signs of infection are visible
3. Persistence — the degree to which a pathogen can fend off its destruction

Whatever methods we choose to treat our excreta, we need to account for the most virulent pathogens with the longest latency and the most persistence. Any and all of our interactions with unsanitized material require precautions so as not to infect ourselves or others (which is surprisingly easy, as discussed in Chapter 3).

Of all the pathogens that exist in the five geo-climatic zones, the most common and persistent pathogen, and the one used as a key indicator for sanitation, is the very persistent *Ascaris lumbricoides* (roundworm). (Buswell et al., 1998; Webber, 2016). *A. lumbricoides* resides as an egg (ova) that requires a lengthy dormancy before it is developed enough to infect a host. This pathogen is the primary target to control for.

Factors influencing pathogen attenuation include:

- **Exit from the body** is the first stage to pathogen death. All organisms have an ecosystem in which they thrive, and when removed from their life support systems, survival is compromised.
- **Time** away from life support systems increases pathogen death. Given long enough with no opportunity for a new host, they will die. Time in hostile conditions speeds up the death rate.
- **Temperature** is extremely effective at physical deactivation — otherwise known as killing the pathogen. Additionally, when this heat is a product of thermophilic processing, there will be more microbes present to chew up and spit one another out.
- **Competition and predation, alluded to directly above,** is another factor that leads to pathogen death. A large and diverse population of organisms will tend to compete for the same foods and see one another as food sources. Yup, they eat each other. When food sources decline, they begin to consume their own structures (protoplasms), which weakens them further. Fancy way to say they starve to death.

- **Antagonism** also promotes pathogen death, via the creation of toxic substances that kill other organisms (some bacteria and fungus create antibiotics). Yup, toxic warfare in the compost pile.
- **pH** both high (alkaline) and low (acidic) will be effective at reducing pathogens. High pH is marginally effective over pH 9 and very effective at pH 12. In mesophilic situations in which pH is raised to over 9.4 and urine is collected with the other raw materials, Ascaris eggs are killed in three months (Jensen, et al., 2009).
- **Ammonia** provides another tool to reduce pathogens. However, if ammonia is used in the process, it should be combined with other methods: research done in 2015 showed that some bacteria are evolving to survive ammonia compounds (Jennings et al., 2015). In composts where urine has been diverted (and so is low in ammonia), longer storage is required to kill Ascaris worm eggs (Jensen et al., 2009).
- **Sunlight** oxidizes and kills microorganisms (Webber, 2016, p. 5; Feachem et al., n.d., p. 79).

Thermophilic digestion is perhaps the lowest-cost process that causes near 100% destruction of pathogens (Feachem et al., 1983). Even the lead researcher of the world's most advanced toilet, winner of the Gate's Foundation "Reinvent the Toilet Challenge" (discussed in Chapter 10), notes that the simple thermophilic compost pile is the simplest, most appropriate and effective technology to sanitize human excreta.

Chemical Pollutants

Chemical pollutants may show up in our compost pile from the food we eat, the pharmaceuticals we take, the air we breathe, or even from the personal care products absorbed into our bodies through our skin. Examples of chemical pollutants are pesticides, plastic residues, endocrine disrupters (like Bisphenol A and phthalates), fragrance chemicals we breath, and pharmaceuticals, including cancer drugs, pain killers, and hormones (birth control pills). Chemical pollutants are a little different from pathogens: they are not disease-causing organisms. Yet the same biological processes that occur in composting to "sanitize" pathogens are also involved in transforming toxic chemical pollutants into less harmful or fully safe substances.

The main process of transformation for these pollutants is *oxidation,* a biological process involving a series of steps in cellular digestion. Oxidation relies on water, enzymes, salts, and acids for extracting energy for cellular use. The process breaks apart chemical bonds in molecules, and swaps hydrogen atoms from one molecule to another, thus permanently transforming chemical pollutants to nontoxic molecules. This occurs in a biologically active compost through natural processes. Advanced oxidative processes (AOP) are used in municipal waste treatment systems to detoxify wastewater streams (an example of an AOP you likely learned in Grade 12 science is the Krebs cycle, which drives cellular respiration).

Achieving Sanitation

All organisms will eventually die a natural death. Death can be sped up by creating a hostile environment that promotes predation, competition, heat, starvation, mutation, or other conditions that cause mortality. Yup, it's all-out war on disease-causing bugs in the compost pile.

A pathogen's ability to survive when excreted from its host is called *persistence.*

One that survives for a long time is to be considered highly persistent; those that die off quickly, soon after excretion, have low persistence. The greater the persistence of a pathogen, the more we have to rely on either harsh conditions or just time itself. It is important that we recognize the role of time, even when we have created an ideal hostile environment.

Sanitization can occur in both thermophilic and non-thermophilic conditions. (Sanitation pathways are shown in Figure 2.17.) Both conditions, though, share a common rule: the sanitization process does not

Fig. 2.16: *Influence of Time and Temperature on selected pathogens in night soil and sludge.* ILLUSTRATION CREDIT: GORD BAIRD, ADAPTED FROM FEACHEM ET AL., 1983, PAGE 79.

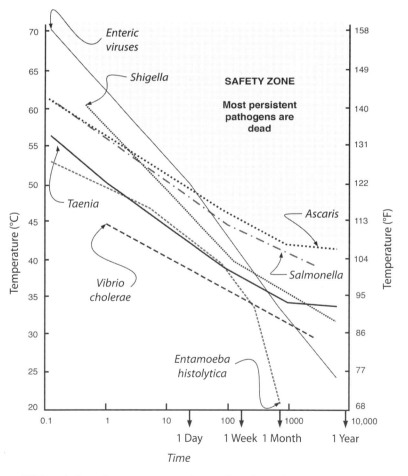

This figure indicates time temperature requirements for pathogen death are at least:
1 hour at > 62°C or 1 day at > 50°C or 1 week at > 46°C

allow for any new additions, hence sanitization only happens as a *batch process* — one "batch" at a time. (*NOTE*: Throughout this book, we will disagree with manufacturers' claims that continuous toilet systems can produce materials that can be immediately buried. These materials may look decomposed, but due to constant additions of raw materials and the corresponding leaching, the levels of volatile nitrogen and pathogens are elevated, which make them unsafe for burial.)

As shown in Figure 2.17, compost sanitation can be achieved by various methods including holding the compost at certain temperatures for certain amounts of time, applying chemical additives, or changing temperatures.

In thermophilic conditions:

Transition to sanitized happens very quickly, requires the least amount of time, and is perhaps the most simple. This essentially is the most common process for the homeowner for this reason. The worst case scenario: procrastinate and let it sit 12 months once temperatures drop.

In non-thermophilic conditions:

Those systems that "molder" can achieve sanitation in several ways. Patience and allowing it to sit for extended times is, of course, the easiest option, but if storage space is limited or nonexistent, then additional options exist: adding external heat to essentially pasteurize the pile; increasing the pH to, in essence, chemically burn the life out of the pile; or boosting the ammonia gas.

Fig. 2.17: *Sanitation pathways for matured materials from thermophilic and non-thermophilic materials sources.*

ILLUSTRATION CREDIT: GORD BAIRD

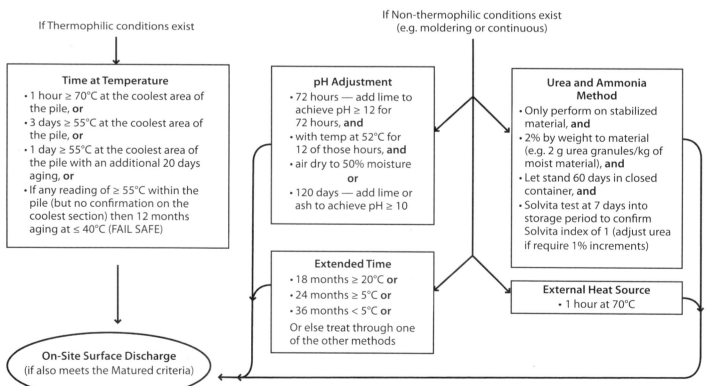

Sanitization Pathways
(batch processing of stabilized materials)

If Thermophilic conditions exist

If Non-thermophilic conditions exist
(e.g. moldering or continuous)

Time at Temperature
- 1 hour ≥ 70°C at the coolest area of the pile, **or**
- 3 days ≥ 55°C at the coolest area of the pile, **or**
- 1 day ≥ 55°C at the coolest area of the pile with an additional 20 days aging, **or**
- If any reading of ≥ 55°C within the pile (but no confirmation on the coolest section) then 12 months aging at ≤ 40°C (FAIL SAFE)

pH Adjustment
- 72 hours — add lime to achieve pH ≥ 12 for 72 hours, **and**
- with temp at 52°C for 12 of those hours, **and**
- air dry to 50% moisture

or

- 120 days — add lime or ash to achieve pH ≥ 10

Urea and Ammonia Method
- Only perform on stabilized material, **and**
- 2% by weight to material (e.g. 2 g urea granules/kg of moist material), **and**
- Let stand 60 days in closed container, **and**
- Solvita test at 7 days into storage period to confirm Solvita index of 1 (adjust urea if require 1% increments)

Extended Time
- 18 months ≥ 20°C **or**
- 24 months ≥ 5°C **or**
- 36 months < 5°C **or**

Or else treat through one of the other methods

External Heat Source
- 1 hour at 70°C

On-Site Surface Discharge
(if also meets the Matured criteria)

- Add urea to induce ammonification (ammonia gas to sanitize).
- Change the pH to chemically burn (denature) protein structures.
- Provide external heating to induce "pasteurization."
- Follow time frames as listed.
- Materials can re-composted in a manner that allows for thermophilic processes to occur; they then need to meet one of the thermophilic pathways, as shown.

Using the most conservative data on pathogen death of the most persistent and virulent pathogens (including *Ascaris lumbricoides*), Figure 2.16 shows the safety zone, representing 100% death. No matter what method of processing you use, so long as it meets the "TIME" at the "TEMPERATURE" curve, and places you in the safety zone, you are assured safety.

Another method that is used as an indicator of sanitization is mentioned by Dr. Jörn Germer. Hardy tomato seeds are deactivated by the same conditions that deactivate parasitic worm eggs (*A. lumbricoides*), and thus the tomato seeds provide an indicator of pathogen death. If the seeds are added at the same time as raw materials, and undergo the same processes, when samples no longer have germinating seeds, then one can be reasonably assured that any potential parasitic worm eggs are destroyed (Germer et al. 2009. "Temperature and Deactivation of Microbial Faecal Indicators.")

Table of Microbiological Risks

Viruses	Death in days at ≥ 55°C (131°F)	Natural death in days at 20°C – 30°C (68 – 86°F)
Polioviruses	14	40
Hepatitis A	14	210
Rotaviruses	14	365
Coxsackie viruses	14	50–270
Adenoviruses	3	60
Norwalk	<1	3
Enteroviruses	14	180
Bacteria		
Salmonella typhi	30 min	60
Salmonella other	1 hr	365
Shigella	1 hr	40
Vibrio cholerae		30
Pathogenic *E. coli*	1 hr	365
Leptospira (spp.)	< 1 hr	< 365
Campylobacter	< 10 min	7
Protazoa		
Entamoeba histolytica	<1	20
Giardia lambia	<3	90
Cryptosporidium	<3	>70
Helminths		
Ascaris lumbricoides Roundworm	1 hr	730
Enterobius vermicularis Pinworm	1	30
Taenia (spp.) Tapeworm	<1 hr	10
Trichuris Whipworm	<1 hr	545
Schistosoma	1	
Necator americanus Hookworm	1 hr	

There are lots of species of Ascaris, but *A. lumbricoides* is the most problematic, most resistant to death, and has the longest latency; therefore, it is an "indicator" pathogen that is used in research.

Compost Quality

Of course, we'd like our finished compost to be of as high a quality as possible. To judge quality, compost can be assessed by testing for its trace mineral element composition (Figure 2.18) and its pathogens and stability (Figure 2.19).

The old adage of GIGO, *garbage-in equals garbage-out,* rings true for what we take into our bodies — from the food we eat to the other toxins we expose ourselves to, from hormones, pharmaceuticals, and body care products. It is possible to test the product of a composting toilet to get a chemical analysis of what went into the toilet. Consider compost testing as akin to a broad-spectrum stool analysis; an opportunity that does not come with your standard flush toilet. If compost testing comes back outside the ranges of Type B Compost in Figure 2.18, then serious consideration needs to be given to changing one's food, water supply, and activities.

Figure 2.19 is the California Department of Resource's Recycling and Recovery's Compost Quality Standards. These qualitative and quantitative measures are used to confirm that a compost is matured, and does not offer harm to the environment.

A fairly new concern has been raised only recently: pharmaceutical contaminants that make their way into the toilet. Many of these chemicals are *not* destroyed by digestion or in regular composting processes, so there is legitimate concern about whether the resultant product is safe. New methods are being developed to investigate pharmaceutical contaminants that are not covered in the standards or compost tests just mentioned. Dr. Gary Andersen, head of the Dept. of Environmental Science, Policy and Management at UC Berkley is doing research in this area. Stay tuned.

Risk aversion by government regulators often overrides scientific facts. All compost regulations and guidelines in North America recommend application only to soils and subsoils around perennial plants; they never recommend application in and around vegetables. We think this is incorrect. If a compost is tested and it exceeds the parameters for maximum concentrations of metals as laid out in Figure 2.18 (Category AA compost) and it also meets the minimum requirements for finished compost quality as laid out in Figure 2.19, then it should be assumed safe for garden use. However, compost testing for the homeowner is probably not feasible. It will be up to the user to determine, but we recommend that if final compost testing is not performed, apply the products only around ornamentals, trees, and shrubs; otherwise, it should be buried and capped with 15 cm (6 in) of soil.

The last thing you want is an immature composter in a mature compost, or perhaps the other way around (Figure 2.20).

Fig. 2.18:

Ontario Compost Quality Standards, 2004, maximum concentrations of metals in compost.

Item	Column 1: Metal	Column 2: Category AA Compost (mg/kg dry weight)	Column 3: Category A Compost (mg/kg dry weight)	Column 4: Category B Compost (mg/kg dry weight)
1	Arsenic	13	13	75
2	Cadmium	3	3	20
3	Chromium	210	210	1060
4	Cobalt	34	34	150
5	Copper	100	400	760
6	Lead	150	150	500
7	Mercury	0.8	0.8	5
8	Molybdenum	5	5	20
9	Nickel	62	62	180
10	Selenium	2	2	14
11	Zinc	500	700	1850

Indicator	Quality Standards for Finished Compost	
Visual	All material is dark brown (black indicates possible burning) Parent material is no longer visible Structure is a mixture of fine and medium size particle and humus crumbs.	
Physical	Moisture: 30–40%, fine texture (all below ⅛" mesh)	
Odor	Smells like rich humus from the forest floor; no ammonia or anaerobic odor.	
Nutrient	Carbon:Nitrogen ratio	<17:1
	Total organic matter	20–35%
	Total nitrogen	1.0–2.0%
	Nitrate nitrogen	250–350 PPM
	Nitrite nitrogen	0 PPM
	Sulfide	0 PPM
	Ammonium	0 or trace
	pH	6.5–8.5
	Cation exchange capacity (CEC)	>60 meq/100g
	Humic acid content	5–15%
	ERGS reading	5,000/15,000 mS/cm
Microbiological	Heterotrophic plate count	$1 \times 10^8 - 1 \times 10^{10}$ CFU/gdw
	Anaerobic plate count	Aerobes:Anaerobes at 10:1
	Yeasts and molds	$1 \times 10^3 - 1 \times 10^5$ CFU/gdw
	Actinomycetes	$1 \times 10^6 - 1 \times 10^8$ CFU/gdw
	Pseudomonads	$1 \times 10^3 - 1 \times 10^6$ CFU/gdw
	Nitrogen fixing bacteria	$1 \times 10^3 - 1 \times 10^6$ CFU/gdw
	Compost maturity	.50% on Maturity Index at dilution rate appropriate for compost application.
	Compost stability	<100 mg O^2/Kg compost dry solids/hour
	E. coli	<3 E. coli/g
	Fecal coliforms	<1000 MPN/g of dry solids
	Salmonella	<3 MPN/4g total solids

Fig. 2.19:
California Compost Quality Standards, California Department of Resources Recycling and Recovery.

Fig. 2.20:
Mature compost/Immature composter.

<small>PHOTO CREDIT: ANN BAIRD</small>

Reference Material:

References for materials and processing states:

Calloway and Margen. 1971. "Variation in Endogenous Nitrogen Excretion."

Czemiel. 2000. "Phosphorus and Nitrogen in Sanitary Systems."

Eastwood. 1973. "Vegetable Fibre."

Feachem et al. 1983. *Sanitation and Disease.*

Goldblith and Wick. 1961. *Analysis Of Human Fecal Components.*

Meinzinger and Oldenburg. 2009. "Characteristics of Source-Separated Household Wastewater Flows."

Rose et al. 2015. "The Characterization of Feces and Urine."

Schouw et al., 2002. "Composition of Human Excreta."

Silvester et al., 1997. "Effect of Meat and Resistant Starch on Fecal Excretion."

Simha and Ganesapillai. 2016. *Ecological Sanitation and Nutrient Recovery.*

Vinnerås. January 2002. *Possibilities for Sustainable Nutrient Recycling.*

Vinnerås et al. 2006. "The Characteristics of Household Wastewater."

References for pathogens:

ENVIS. 0ctober 2016. *Pathogens in Human Excreta.*

Feachem et al. 1983. *Sanitation and Disease.*

Food and Agriculture Organization, OIE — World Organisation for Animal Health, World Health Organization, and Department of Food Safety, 2006.

References for pathogen groups and their lifespans:

Alberts et al. 2002. "Introduction to Pathogens."

Buswell et al. 1998. "Extended Survival and Persistence of *Campylobacter.*"

Betancourt, W.Q., and L.M. Shulman. 2015. "Polioviruses and other Enteroviruses."

ENVIS. 2016. "Pathogens in Human Excreta."

Katayama, V. 2015. "Norovirus and other Caliciviruses."

Sobrados-Bernardos, L., and Smith, J. E. 2012. "Controlling Pathogens and Stabilizing Sludge/Biosolids."

Sossou et al. 2013. "Inactivation mechanisms of pathogenic bacteria."

Stanford University/ParaSites. 2012. The Mentor Initiative: Stanford University/ ParaSites 2012. *Index of Group Parasites.*

References for compost quality:

Carballa et al. 2008. "Comparison of predicted and measured concentrations of selected pharmaceuticals, fragrances and hormones in Spanish sewage."

Schröder et al. 2016. "Status of hormones and painkillers."

Ternes and Joss (eds.) 2008. *Human Pharmaceuticals.*

Wichuk and McCartney. 2010. "Compost Stability and Maturity."

Chapter 3

System Components and Processes

The Components of a System

GENERALLY, a compost toilet system has five components:

1. **Toilet pedestal** — Where the action happens.

2. **Conveyance** — The route materials take to get to the collection vessel.

3. **Collection** — The intermediate stage and container where materials are collected before final processing.

4. **Processor** — Where the majority of biological processes occur to *fully* decompose and/or sanitize materials. (A system can rely on several processors.)

5. **Fluid Management** — Some systems do not require fluid management, but for those that do, it is a critical feature (see Chapter 8).

Hygienic Design

The multi-barrier approach introduced in the last chapter and represented in the F Diagram (Figure 2.15) relies on all components (and systems) functioning properly to limit pathogen transfer. Good design ensures that components are easily accessible, durable, and replaceable; the user should be able to clean and service all components without any yuck factor. It's also critical that at no time are people at risk.

We are exposed on a regular basis to pathogens from wiping our own parts, dealing with toilet overflows, baby diapers (they can be toxic), our pets' feces, and animal manures. We rely on good habits and judgement when encountering these; we wash our hands, keep toilet seats and bowls clean, and clean up messes when required. With compost processors, collection and transport containers, cleaning implements, and tools like hoses, rakes, shovels, and thermometers, it is important to employ the same good habits. This means keeping items dedicated for the sole purpose of servicing our systems separate from all other tools (we find it useful to keep our compost equipment in the enclosed fenced area that contains the compost pile). Wherever you store your equipment, be sure everything is labeled. You might need some signage as well. It should be absolutely clear to anyone coming upon your tools that some are *only* to be used for the composting toilet system.

The cleaner we are, the less attractants there will be. What are attractants? They are any items that another pesky life form considers as food; spillage and remnants of materials in an area where they are not supposed to be will attract critters. Clean and tidy is the first rule.

Hand washing

Hand-washing facilities provide the first physical barrier for pathogen transfer. This is the most important and effective barrier — which is why it is also used in the medical profession, the food industry, and as a part of public health initiatives during virus outbreaks. Ensure that hand-washing devices (hand sanitizers, or hot water/soap facilities) are readily available in the bathroom and near all maintenance and servicing processes.

Ventilation

Ventilation is critical for controlling moisture and odor. Once you have had the pleasure of using a vented compost toilet, there is no going back to a stinky flush toilet or a stinky bathroom. A proper ventilation system will also control for insects and other pests. Good venting design will eradicate downdrafts, condensation in the pipes, and excessive noise; it must also ensure that the wiring is safe, and that rain can't enter the system.

Other household mechanical systems should be configured so they do not impact proper venting. The rest of this section examines the key design considerations for ventilation.

Vent pipes

- **Size** — Pipe will range in size from 3"–6" (90 mm–170 mm); the size you need will depend on the fan you choose. Most homemade systems will have 4" (110 mm) plastic pipe.

Fig. 3.1: *Ventilation: Air exit options.* ILLUSTRATION CREDIT: GORD BAIRD, ADAPTED FROM THE LITTLE HOUSE COMPANY, WWW.LITTLEHOUSE.CO

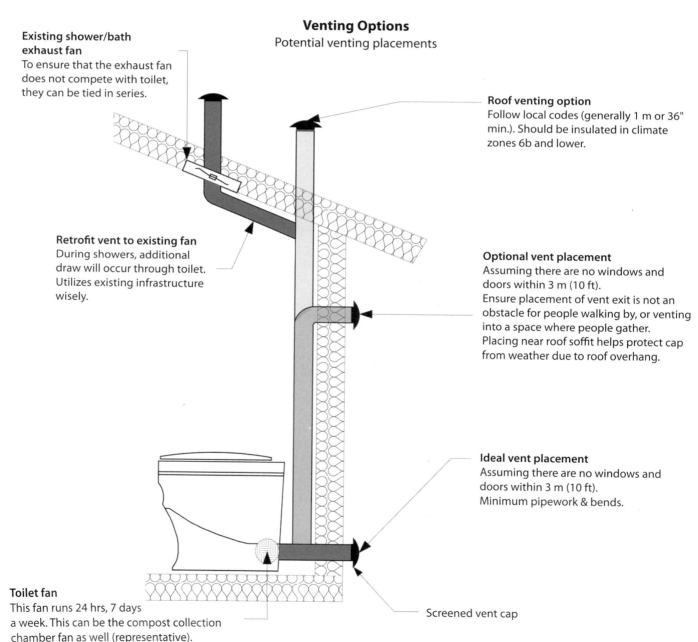

Venting Options
Potential venting placements

Existing shower/bath exhaust fan
To ensure that the exhaust fan does not compete with toilet, they can be tied in series.

Roof venting option
Follow local codes (generally 1 m or 36" min.). Should be insulated in climate zones 6b and lower.

Retrofit vent to existing fan
During showers, additional draw will occur through toilet. Utilizes existing infrastructure wisely.

Optional vent placement
Assuming there are no windows and doors within 3 m (10 ft).
Ensure placement of vent exit is not an obstacle for people walking by, or venting into a space where people gather.
Placing near roof soffit helps protect cap from weather due to roof overhang.

Ideal vent placement
Assuming there are no windows and doors within 3 m (10 ft).
Minimum pipework & bends.

Toilet fan
This fan runs 24 hrs, 7 days a week. This can be the compost collection chamber fan as well (representative).

Screened vent cap

- **Slope** — Good plumbing practice should never have a vent pipe travel horizontally in unheated cavities because moisture will condense inside the pipe and potentially pool, offering habitat for insects and mold. Where a horizontal run is unavoidable, it may be necessary to install a condensation drain — a small drain port using ¼" tubing, which is plumbed and drained to daylight.
- **Insulation** — Vent pipes traveling through unheated cavities should be insulated to stop condensation. Exterior vent pipe-runs should be insulated; in many cases, they should be boxed in.
- **Exits** — Vent pipe exits should not terminate within 3 meters (10 ft) of an opening window or door; this helps to prevent odors from being drawn into the building. Preferably, the vent pipe terminates through the roof in the same format as is normal practice for other vent pipes. Options are shown in Figure 3.1.
- **Caps/Screens** — Caps and screens help to prevent cross-flow winds from back-drafting down the pipe; they also exclude insects and other pests from entering. The screen needs to be accessible for cleaning at least once a year because lint will build up and impede airflow. Screens made of a 16 or 18 mesh are suggested.

Vent fans

There are two general types of fans that are used: *centrifugal fans* and *axial fans*

Centrifugal fans Also called "blower" or "squirrel" fans, these are the most efficient; they suck air in by impellers and displace it radially, resulting in a high air flow and higher pressure — more than is required for small toilet systems, so these tend to be used only on larger multi-pedestal systems. The best search terms to use for finding these

fans are *centrifugal* and *inline*. Manufacturers include Fantech (FR Centrif models) and Soler & Palau (TD and PRF models).

Axial fans are generally the cheapest, most commonly used, and easiest to source (see Figure 3.2). Air flow in these travels parallel to the shaft; this is a less efficient design than the centrifugal fans, because the shaft itself is an obstacle that impedes air flow. These fans are noted for having a low pressure air flow. Here are some examples, both AC and DC:

- Orion Fan — 80mm × 25mm; 115 VAC, 4W; 20 dBa; 13cfm; Prt# OA825AP-11–3WB
- Orion Fan — 80mm × 38mm; 115 VAC, 5W; 18 dBa; 17cfm; Prt# 0A80AP-11–3WB
- Qualtek — 80mm × 25mm; 12 VDC, 600mW; 16.5 dBa; 20 cfm; Prt# FAD1–08025CSAW11
- ebm-papst — 80mm × 25mm; 24VDC, 800mW; 17 dBa; 19.4cfm; Prt# 8414NLU
- NMB Technologies — 60mm × 25mm; 24VDC, 720mW; 17 dBa; 14.5cfm; Prt# 2410SB-05W-B20-B00

No power for a fan? Although we strongly advise the use of mechanical ventilation, we realize this may not be always be possible. For alternatives, check out the USDA's "Sweet Smelling Toilet Installation Guide" (Land, 2003).

Fig. 3.2: *Axial fan.*

Competing air flows

In well-sealed rooms or homes, a draw of air supply (e.g. stove fans) can outcompete the compost toilet ventilation fan and draw odor out of the toilet system. Not good. Other home systems that draw air include HRV, HVAC, wood fireplaces, and bathroom exhaust fans.

Make-up air — One simple solution that avoids sucking air out of the toilet is allowing extra air inputs into the room through windows or external air inlet vents. This ensures that the fan is pulling air in from somewhere other than the toilet.

Fan power — Increased fan power (a higher velocity flow rate) can overcome competition, but this is not necessarily efficient; the fan at the toilet has to work harder than necessary, so it consumes excess power.

Combined ventilation — As shown in Figure 3.1, a competing bathroom fan can be used to the benefit of the toilet system if the two are tied together. As shown in Figure 3.3, you can use the existing duct work and roof exit that is generally inaccessible, yet completely functional; this avoids having to perform costly renovations to install additional venting and roof penetrations. When the bathroom fan is not in use, the toilet fan performs its regular function; when the bathroom fan is turned on, additional air is drawn through the toilet seat, pulling with it the moisture from bathing and showering.

Building code air change — In the building codes in the past decades, there has been a move toward creating more tightly sealed homes. With this has (ironically) come the requirement for additional ventilation. Some, but not all, jurisdictions may allow the toilet ventilation to contribute to the required air change per hour (ACH), which is usually 0.3 (30%). If this approach is approved by the building inspector, the following calculations would be used.

Air Change Calculation:

Volume of the building = length × width × height

Required ACH = 30% × volume

Expected fan ACH = predicted fan cfm × 60 minutes

Example: *House that is 30′ × 40′ with 9′ ceilings:*

Volume of the building = 30 × 40 × 9 = 10,800 ft³

Required ACH = 30% × 10,800 = 3,240 ft³

The house has two toilets, each with its own fan:

Two 15 cfm fans = 30 cfm

Volume of air change from toilets = 30 cfm × 60 minutes = 1,800 ft³

1,800 ft³ ÷ 3,240 ft³ =0.55

Therefore, the toilet system would provide 55% of the required ACH.

Fig. 3.3: *Compost toilet retrofit using existing ceiling fan as ventilation system.*

Photo credit: Gord Baird

Tricks and tips

Vibration — Fans may cause plastic vent pipes to vibrate with unpleasant magnified acoustics; therefore, when installing the fan assembly in line with the vent pipe, attach the assembly with either duct tape or rubber couplers to stop direct contact of the fan with the rigid pipe work. If you can choose between two fans that perform the same job, choose the one with the lower RPM; this is the one likely to have the lower decibel (dBa) sound rating.

Assembly — To make a fan assembly from scratch to house an axial fan, two toilet flanges can be used to sandwich the axial fan. This fan assembly unit is then attached in line with the vent work via two rubber couplers (or duct tape) which act as a vibration isolator. Figure 3.4 illustrates the configuration.

Placement — For best airflow, place the fan as close to the exit as possible while still allowing easy access for cleaning. Fans are more efficient at sucking (or pulling) air than blowing (or pushing) it through the vent pipe.

Cleaning — Place a junction box within the vicinity of the fan assembly, and make the assembly easy to access for cleaning. To clean, remove the fan and wipe it with warm water, dry it, and then reinstall. Silicone sprays used for lubricating can help to reduce the amount of dust particles adhering to the fan blades.

Fig. 3.4: *Fan Assembly Detail: Toilet flanges, axial fan, and flexible coupling are used here.*

ILLUSTRATION CREDIT: GORD BAIRD

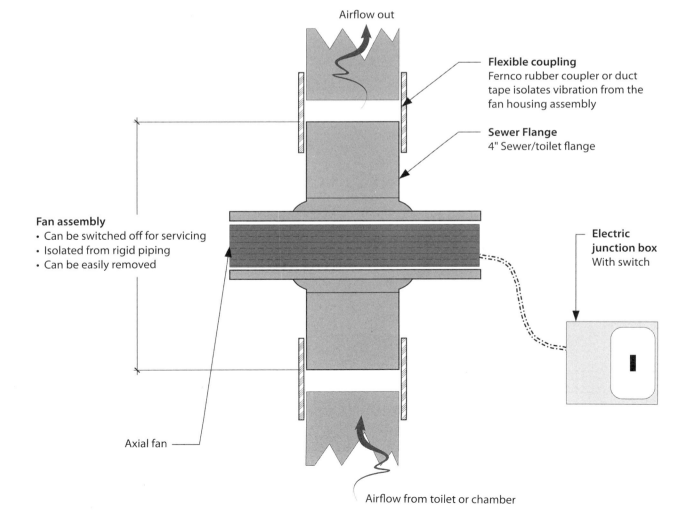

Airflow out

Flexible coupling
Fernco rubber coupler or duct tape isolates vibration from the fan housing assembly

Sewer Flange
4" Sewer/toilet flange

Fan assembly
• Can be switched off for servicing
• Isolated from rigid piping
• Can be easily removed

Electric junction box
With switch

Axial fan

Airflow from toilet or chamber

In our *Eco-Sense* home, the ventilation for the toilets was permitted to be sized to meet the code requirements for air changes per hour (ACH). The fans exhaust enough air through the toilet/collection systems to avoid the need for additional HVAC or HRV mechanical systems. In addition, the fans serve as the humidity removal device for the bathrooms and showers by pulling vapor *down* into and through the toilet, and then exhausting it out of the home. The concept of a bathroom fan in the ceiling is commonplace, but it has the impact of removing the warmest air from the highest point in the room. Placing the fan lower down allows the warmer air to remain in the room (thus reducing heat loss) *and* the water vapor can still be removed (this idea came from Larry Schlussler of Sun Frost [Schlussler, 2002]).

Vent pipe connections

Vent piping has to interface with the toilet system somewhere, and that could be from a toilet pedestal, a toilet cabinet, a storage tank/chamber, or a storage room. Wherever the ventilation is placed to best serve the design, the goal is to vent the fumes outside quietly and securely, meaning connections must be strong and resistant to vibration, bumps, and thumps, and they must be able to handle the shifting that occurs from heating/cooling. We do not want shifting to break the seal. If the seal is lost, unpleasant odors may enter the indoor environment. Figure 3.5 illustrates three of the four seal options discussed below.

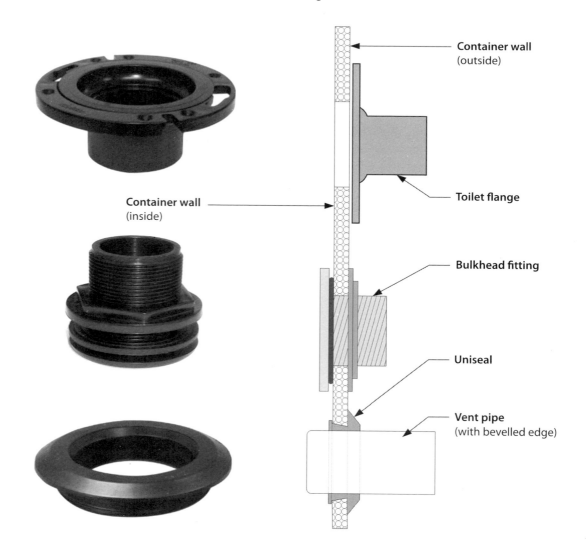

Fig. 3.5: *Vent pipe connection options: toilet flange, bulkhead fitting, and uniseal.*

ILLUSTRATION CREDIT: GORD BAIRD

Toilet Flange — The same toilet flange used to make the fan assembly can be used as the attachment point for vent work. The flange is screwed or bolted to the item you want to ventilate.

Bulkhead Fitting — Bulkhead fittings are the most robust method. They are a two-piece unit that threads together and pinches the wall of whatever you are working with between the two halves, creating a water-proof connection.

Uniseal — Loved by some and hated by others, the uniseal offers another option. The uniseal is O-ring-like, designed to be inserted into a drilled hole, and then have pipework inserted through it. The insertion of the pipe expands the O-ring, creating a watertight seal for low-pressure applications. Four tricks when using them:

- Drill the exact specified hole for the uniseal.
- Bevel the outer edge of the pipe that will be inserted.
- Lubricate with something slippery.
- Swear when you push the uniseal through the hole into the chamber.

Polyurethane Sealer — This option is the one of last resort if you either do not have access to a mechanical system noted above, or the confines of the space don't allow their use. If you go this route, choose a high-quality marine-grade polyurethane sealer over a standard carpenter's contractor grade glue/sealer. If you have to use a sealer, do not use silicone as an alternative to poly-urethane; it is less toxic, but it also has poor long-term adhesion to a variety of materials,

lending to failure sometime in the future. We recommend SikaFlex 291, but be sure to use it in well-ventilated area, as it off-gasses dangerous compounds until cured.

Pedestal–User Interface

The toilet pedestal or toilet cabinet (the thing you sit on to do your business) is, in essence, the user interface. It is comprised of the toilet seat and the seat's supporting structure. Toilet pedestals come in a variety of forms: some collect everything (urine, feces, and toilet paper), some divert urine at the seat, others involve micro-flushing with water or foam, and a few are full flush, like a standard toilet.

Whatever the type, the structure needs to be of sufficient physical strength to support a large weight (figure for about 180 kg [400 lbs]) and finished with a coating that will withstand regular cleaning and scrubbing, resist scratching (belt buckles, sand and grit), and chemical etching from various cleaners (baking soda, vinegar, peroxide, bleach, and various toxic and obnoxious cleaners — not that anyone would use these…right?). In our home, we clean with a soft cloth and soapy water or vinegar.

Manufacturers often use ceramics, high-density plastics, or gel-coated fiberglass cabinets. Site-made (homemade) toilets can be made of plywood coated appropriately with epoxies, fiberglass, or finishes designed for wooden floors (e.g. urethane-based fin-ishes containing aluminum oxide). Examples of different pedestals are shown in Figures 3.6 through 3.10.

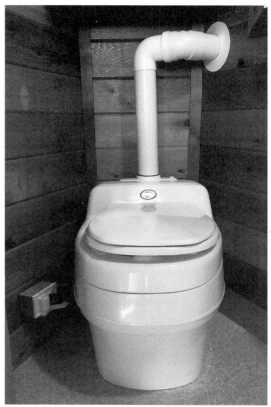

Fig. 3.6 (top left):
Commode batch toilet pedestal with ¾" marine-grade plywood finished with an aluminum oxide polyurethane floor finish; standard toilet seat.
Photo credit: Elisa Rathje

Fig. 3.7 (top right):
Separett plastic toilet pedestal.
Photo credit: Gord Baird

Fig. 3.8 (bottom):
Gravity-drop ceramic pedestal with urine separation by Ecosan.
Photo credit: SuSanA
(Sustainable Sanitation Alliance)

Fig. 3.10 (above): *Jets' Jade ceramic vacuum flush pedestal.* Photo credit: Jets Vacuum AS, Hareid, Norway

Fig. 3.9 (left): *Fiberglass gel coat pedestal/cabinet for a commode batch compost system at a public park.* Photo credit: Gord Baird

Conveyance

Conveyance is the route that the raw materials travel from the toilet seat to the initial place of collection; conveyance is very different for different systems. Full details for specific systems will be discussed in future chapters.

For a commode batch system, conveyance is nothing more than the air space between the human body and the bucket below. Some of the larger collection chamber systems will have a drop chute 10" to 12" in diameter that direct the raw materials. Some systems that serve multiple toilets from pedestals on various floors will incorporate flush systems that use foam, micro-flush water, or vacuum pumps. Figure 3.11 pictures four conveyance methods and how they can be situated.

Direct drop

Pretty much no description required; just gravity, no obstacles. If it is a direct drop into a container, then there is no chute cleaning. Any questions? But what if you have to use a chute as part of a direct drop? You will have

Fig. 3.11: *Conveyance applications for micro-flush, direct drop, and Jet flush toilet systems; adapted from Cape Cod Toilet Center.*

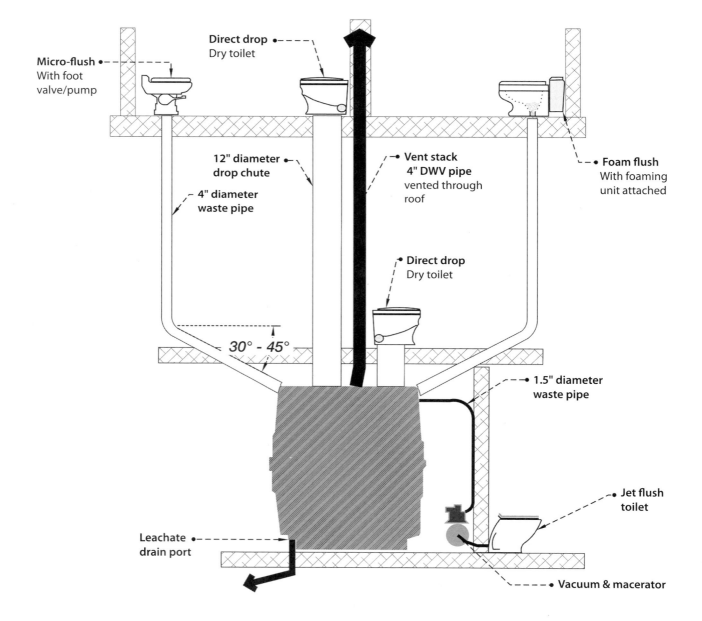

Micro-flush • — —
With foot
valve/pump

Direct drop • — — —
Dry toilet

12" diameter • — —
drop chute

⌐ 4" diameter
waste pipe

Vent stack
4" DWV pipe
vented through
roof

Foam flush
With foaming
unit attached

30° - 45°

Direct drop
Dry toilet

1.5" diameter
waste pipe

Jet flush
toilet

Leachate • — — —
drain port

Vacuum & macerator

poo hit the sides, and you will be cleaning on a weekly basis. (Thus, using chutes longer than a couple feet is not our preference.)

Vacuum flush

Vacuum-flush conveyance systems use suction to transport excreta from a specialized pedestal through a narrow pipe to various types of collection systems; they usually employ between 0.2 L to 1.5 L (7 to 51 ounces) of water. There are two versions of vacuum toilets: *Constant Vacuum Systems (CVS)* that maintain a constant vacuum in the pipe, and *Vacuum on Demand (VOD)* systems, in which vacuum is created when the unit is flushed. Both systems rely on a specialized pedestal with a sealing valve; when a button is pushed, the valve opens, and materials move into the pipe as a *slug*, basically being pushed by the outside air rushing in to fill the vacuum. This slug is drawn through a macerator before being deposited in the collection bin. Sound effects with these systems can be quite entertaining.

CVS systems are used in large structures with lots of toilets, and they require constant power. Thus, they are not generally used in residences.

VOD systems are more commonly used in residential applications. They easily accommodate multiple toilets. The vacuum pump only turns on when a button is pushed (see Figures 3.12 and 3.13). Cycling on/off power usage of short duration combined with these systems' availability in both DC and AC voltage options makes them surprisingly well suited for cabins, off-grid homes, and small boats, where they are often found.

The technology is expensive for VOD systems ($4,000+ for the pedestal and pump), but if faced with no choice but to pump horizontally (or up, vertically), VOD is the only option. However, if you compare it with the cost of a regular flush toilet when a septic system is required, the vacuum flush will win out. Plus, they are known for their reliability, so the technology should not be a detractor.

Vacuum systems offer great flexibility in retrofitting existing buildings. Odor is automatically controlled at the pedestal due to the presence of the vacuum valve in

Fig. 3.12: *Jets vacuum on-demand standard system.*
Photo credit: Jets Vacuum AS, Hareid, Norway

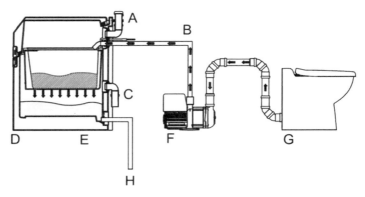

A) Ventilation outlet
B) Inlet pipe
C) Ventilation inlet
D) Discharge outlet
E) Biotank
F) Vacuumarator™ pump
G) Jets™ toilet
H) Filtration trench, sewer, use as fertilizer or empty to suitable source.

Fig. 3.13: *Jets 24 VDC vacuum pump: The Ultima C200 Vacuumarator.* Photo credit: Jets Vacuum AS, Hareid, Norway

the toilet, so ventilation is only needed to service the collection chamber, and it is not influenced or impacted by other household ventilation systems.

As noted, these systems are versatile. They can be used in many different situations and configurations, from holding tanks in boats to self-contained compost chambers, like those we will discuss in Chapter 7.

On the down side, misuse (deposits of sanitary napkins, condoms, dropped toys or dentures) will clog the pump and cause a blockage.

Micro-flush

Micro-flush units are used when toilet pedestals are not aligned directly above the compost collection chamber; in this situation, the solids need a little help to get to their destination. Micro-flush can employ either water or a foam medium consisting of water and a biodegradable soap. The key point is that the pipework needs to be either vertical or on a steep slope (30°–45°). Pipe diameter will be specified by the manufacturer of the unit. If the diameter of the pipe is too small or large, or if the pipe slope is too shallow or steep (other than vertical), or if there are too many directional changes, or if the plumber has not paid attention to detail and left burred edges on the inside of cut pipes, conveyance will not happen smoothly, and blockages will occur. The system may get constipated. So, you must pay attention to details.

The foam generators can be separate stand-alone tanks that are plumbed from a central location, or they can be a small unit attached to the toilet. Figure 3.11 shows a schematic of a micro-flush foam toilet with integrated unit, as would be found with a Nepon or Neptune pedestal.

Mechanical conveyor

Mechanical conveyors, as pictured in Figure 3.14, are used to transport solids up and

Some of the information in this section comes from Stauffer, B. (n.d.) "Sustainable Sanitation and Water Management Toolbox: Vacuum Toilet." Check it out if you are considering this option.

Fig. 3.14: *Mechanical conveyance system of the Ecodomeo toilet (with exterior shell removed), allows solids to transfer up and away, while allowing fluids to drain to a dedicated leachate system.* Photo credit: Emmanuel Morin, Ecodomeo, www.ecodomeo.com

away to a storage area, while allowing fluids to drain down and away. These systems are connected to a leachate handling system, so fluid *collection* is not part of these systems (and thus, resource recovery of the nitrogen-rich fluids is lost). That said, with mechanical conveyors, the solids can be moved without electricity via a mechanical foot pedal and then stored and decomposed in an area not requiring drainage.

Collection

Collection containers come in various forms — from simple containers to much more complex chambers that may whiz, whir, spray, spin, rotate, roll, heat, and scrape. So many options. The following systems are explained in greater detail in Chapters 5 to 7.

Commode batch: bucket/bin

Pictures are worth a thousand words. Figure 3.15 shows how small containers, such as 5-gallon buckets, can be used. Figure 3.16 shows an example of one of the manufactured systems that employ a bucket lined with a decomposable cellophane bag. This is what's inside the toilet shown in Figure 3.7 (made by Separett).

Commode batch systems are covered in more detail in Chapter 5.

Batch/moldering: bins/chambers

Larger bins are also used in a variety of systems including the moldering system and upsized commode batch systems. Collection containers range from 80 L to 1,000 L. Generally, these are used where it is desirable to collect materials over longer periods of time, and where, in many cases, an initial degree of the processing is undertaken as materials rest within the

Fig. 3.15: *Bucket collection for a commode batch using black HDPE 5-gallon buckets.*
Photo credit: Gord Baird

Fig. 3.16: *A biodegradable cellophane bag used in a Separett commode batch Villa 9000 system.*
Photo credit: Separett

container once filled. (See Figure 3.17 and Figure 3.18.)

Systems that have bin/chamber collection are covered in more detail in Chapter 6.

Chamber Collection: Continuous

Chamber collection comes in several forms, but all are characterized by allowing a certain level of decomposition processing to occur at multiple stages: from receiving new additions of raw excreta through to a stabilized decomposed product (thus, the term "continuous"). These systems can be small one-piece self-contained or larger two-piece centralized systems.

Continuous systems are covered in more detail in Chapter 7.

One-Piece Self-Contained Systems

One-piece, single-unit sealed systems are stand-alone units in which the toilet pedestal and collection chamber are one unit. This allows installation anywhere at any time, with little to no modification of an existing structure. These systems come complete with venting ports, fans, devices for churning or turning the composting materials, and an access drawer for removal of contents; some have heating elements to aid processing

Fig. 3.17: *Public washroom facility using 80 L green organics for a commode batch system.* Photo credit: Gord Baird

Figure 3.18: *Moldering toilet system at the Cape Cod Eco-Toilet Center using 64-gallon wheeled collection bins.* Photo credit: Earle Barnhart, Cape Cod Eco-Toilet Center

and evaporation of excess moisture, and some have a leachate drain to allow fluids to drain. These self-contained units are usually designed for occasional-use situations, like a summer cabin or an RV. However, as seen in Figure 3.19, these systems easily integrate into a bathroom without looking temporary or provisional.

Two-Piece Centralized Systems

Two-piece systems are characterized by a toilet pedestal that conveys feces to a larger chamber below. The larger storage chamber offers the advantage of requiring less

servicing. In fact, a big chamber can receive and hold materials for several years. Figure 3.20 shows a large chamber designed to service a four-person household, often referred to as an *inclined-plane* or *Clivus-style.*

Moisture Management

For continuous and chamber batch (moldering) systems, urine management is critical; too much moisture leads to an anaerobic septic soup that breeds vermin (flies), creates gasses and smells, and halts the composting process. Urine diversion reduces the demands placed on the other moisture management

Fig. 3.19: *Sun-Mar Excel self-contained continuous compost toilet.*
PHOTO CREDIT: R. DAVEY, 2009, CREATIVE COMMONS

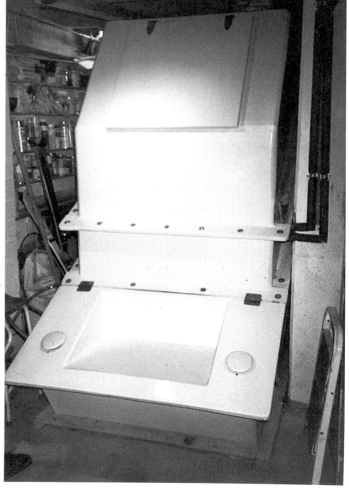

Fig. 3.20: *Clivus-like insulated container for four-person household at the ecological settlement in Hamburg-Allermöhe, Germany.*
PHOTO CREDIT: WOLFGANG BERGER, 2004

Urine Diversion — to Separate or Not?

A topic that is bound to raise debate is whether or not to separate urine in the collection process. There is no right or wrong answer. We encourage you to consider the points raised in this chapter and find a solution that best meets your goals and the practical implementation of the best system for you.

There are four compelling reasons for diverting urine:

1. To use it as a fertilizer (pure or diluted, after sanitization)
2. To reduce odor in systems that employ large collection chambers
3. To reduce liquid loading that creates anaerobic conditions, slows decomposition, and encourages flies
4. To limit the addition of bulking agent and thus limit the rate at which a collection chamber fills

For commode batch (Humanure) systems, collection of everything (*all-in*) is preferable. This is partly because servicing tends to be on an as-needed basis, so it is irregular. But it's also true that the bins can be emptied directly into compost piles, which tend to need the moisture, anyway.

Generally speaking, continuous and chamber batch (moldering) systems will most likely have a form of urine diversion to meet reasons 2, 3, and 4, while commode batch system will only divert for reasons 1 and 4.

tools and reduces the sizing of the chambers required. Though bulking agents can be used to absorb excess fluids and offer up a carbon source to balance the nitrogen, relying on them for absorption requires collection units sized three times larger than if they were just collecting the excreta. With practicality as our guide, it makes more sense to control moisture by evaporating and draining it using fans/ventilation, heating elements, and leachate drains. Urine diversion at the pedestal greatly enhances the ability of these moisture management tools to perform their job.

If you are focused on nutrient collection for use as fertilizer, see the discussion about urine collection in Chapter 8.

When is it wise to consider urine diversion?

- in high-use facilities
- in moldering toilet systems
- in remote locations relying on small collection chambers and infrequent servicing
- in locations where there is a lack of access to bulking-agent materials
- for small properties where zoning setbacks don't allow for acceptable compost pile placement
- in large centralized continuous collection systems
- if vermicomposting is incorporated into initial decomposition. (You don't want to drown the worms!)

Clearly there are many good reasons for considering urine diversion, but there are also some important drawbacks.

The biggest disadvantage is that diverting urine means a lot less nitrogen in the initial compost. This results in reduced microorganism growth and slower decomposition. The low level of microbial activity results in less internal heat production and longer time periods before materials reach their desired state. This may require the movement of less processed materials to composting piles for further decomposition.

The second disadvantage comes from the urine-diverting toilet seat design that enables separation. Though many see the idea as favorable and a concept they think they can get used to, many find that using such systems is not always practical (Lienert and Larsen, n.d.; Toilet Revolution, 2016). Research shows that despite there being conceptual acceptance of urine diversion, habits are hard to break (Rieck, von Münch, and Hoffmann, 2013). Many toilet systems with urine-separating seats require signage and stickers showing "how to use the toilet." Think about this in the context of potty training young children or having to explain what's going on to the visiting in-laws. The conversations would not only have to cover differences in male and female anatomy, but also the topic of how to control both pee and poo (happening potentially at the same time but going in different directions).

A third disadvantage, fits within the "yucky" realm. When accidents happen and feces ends up in places that it shouldn't, the result is an unpleasant maintenance task. The old saying "shit happens" takes on new meaning.

A final disadvantage arises from the composition of urine. Remember that conversation about all the minerals in urine? Well, the drains are prone to scale buildup from these minerals. Even with good design and proper use, the system has to be set up in a way that allows for dismantling and cleaning — using either mechanical methods or chemical methods utilizing acids (Udert, Larsen, and Gujer, 2003), or by replacement of specialized urine traps and chemical/enzyme treatment pucks.

To summarize, urine diversion can:

- reduce the infrastructure required to store materials and manage moisture, leading to volume reduction, and, hence, longer periods between servicing a system
- offer the potential for nutrient collection
- reduce odor and flies
- result in longer composting times
- result in greater need for cleaning and servicing of diverters
- be problematic for untrained users

Chapter 4 will explore the sizing of urine collection systems; Chapter 8 will explore moisture management in more detail.

All-in Collection: Urine and Feces

With the collection of all excreta (urine and feces), there comes benefits in the composting processes. Urine serves two major functions: it provides moisture needed for the biological processes to occur; and it provides the nitrogen source that enables thermophilic bacterial growth, which prompts rapid carbon breakdown. Without urine, the carbon-nitrogen balance (C:N ratio) is not ideal for the support of microorganism growth. Carbon is the energy source and building block for cell tissues, and nitrogen is implicated in the enzymatic processes and is a component of proteins' nucleic acids and amino acids used in cell growth. An imbalance of the 30:1 ratio will impair life in the compost (Trautmann, Richard, and Krasny, 2017).

Batch commode collection systems deal with this "all-in" approach easily because they can be serviced (dumped and processed) as much or as little as needed. Excess moisture is not an issue because bulking agents are added to absorb the excess moisture, and in doing so also they provide the added carbon to balance the nitrogen. When everything is deposited into the composting pile, all ingredients are present to enable rapid natural thermophilic processes to begin.

The other systems — chamber batch (moldering) and continuous — need robust drainage design if all fluids are collected; leachate drain systems must be added to ensure there is no excess buildup of moisture. If part of this design fails (breakage, blockages, or power outage), the system fails.

The Yuck Factor

Let's face it, the yuckier something is, the less likely we are to deal with it. If we avoid regular maintenance, we increase the risks of failure of the system, we increase the health risks, and we increase the yuck factor. The best systems are the ones that are maintained — and it is usually the simplest that are the easiest to maintain. Ultimately, no matter which system you use, you will have to look at raw excreta at some point.

Compost Processor

No matter which compost toilet you choose, if your desired outcome is to process your resources to the highest level for the safest re-introduction to the landscape, then the ideal final stage in every process is to batch compost the materials through a thermophilic stage. The next best is to at least batch process

them in a mesophilic compost processor for the required extended timeframes. What that means is that at some point, to achieve a safe and sanitized compost, materials will need to be set aside, one way or another, and given some form of treatment. The most convenient place for this is the compost pile.

It is in the compost pile where the most action happens. However, the compost pile usually receives the least attention and diligence in design. It is the only component that houses any true composting activity; and, outside of using chemicals, added heat, or extensive retention times, it is the only component where full sanitization occurs within short timeframes.

Design recommendations

Design for the "4S's": *safety, sanitation, signage,* and *security.* Figures 3.21 and 3.22

Fig. 3.21: *Double-bin compost processing pile considerations.*

Illustration credit: Gord Baird

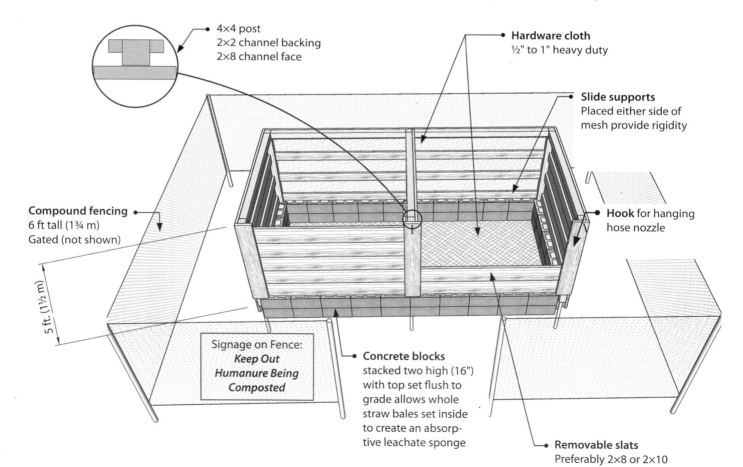

- **4×4 post**
  2×2 channel backing
  2×8 channel face
- **Hardware cloth**
  ½" to 1" heavy duty
- **Slide supports**
  Placed either side of mesh provide rigidity
- **Compound fencing**
  6 ft tall (1¾ m)
  Gated (not shown)
- 5 ft. (1½ m)
- **Hook** for hanging hose nozzle
- Signage on Fence:
  Keep Out
  Humanure Being
  Composted
- **Concrete blocks**
  stacked two high (16")
  with top set flush to grade allows whole straw bales set inside to create an absorptive leachate sponge
- **Removable slats**
  Preferably 2×8 or 2×10

provide the visual reference for the following design recommendations for a two-pile system. (This can be expanded to a three-pile system.)

There are seven key factors to be considered in compost processor (or pile) design. Here are the seven items, along with considerations for each:

1. **Durability:**
   - The sides or walls must be built so as not to rot out or splay apart when filled.
   - Appropriate materials need to be used: concrete cinder block, cedar, and galvanized lawn fencing material with a minimum of 12-gauge wire.
   - Packing pallets don't cut it. They need to be replaced too often.

2. **Serviceability:**
   - Easy access and room to maneuver should be built into the design.
   - Adjustable height and/or lid is required.
   - You should design in a movable step or standing platform to allow easy access as the compost pile increases in height.
   - The height of the piles should be limited to 1.5 m (5 ft).
   - Fenced area in front should have opening gates for easy handling of rakes, wheelbarrows, and collection container movement.
   - Composting area should be located within easy access to the home and to a water source.

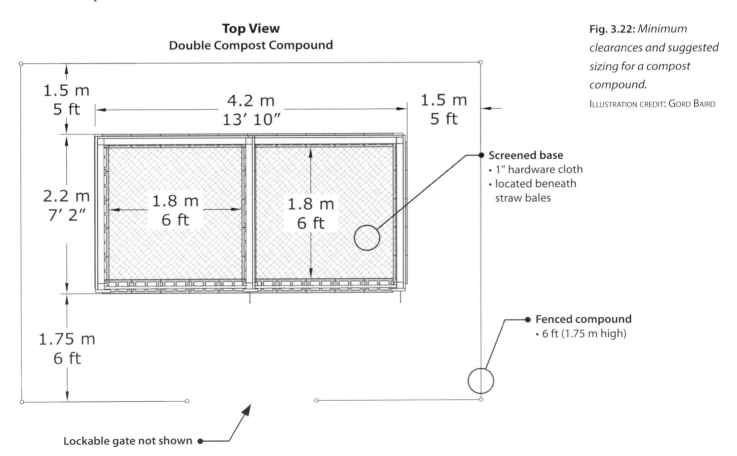

Top View
Double Compost Compound

Fig. 3.22: *Minimum clearances and suggested sizing for a compost compound.*
ILLUSTRATION CREDIT: GORD BAIRD

3. **Absorptive Mat:**
   - This sponge-like mat absorbs leached moisture in the early stages of the pile development — until the compost develops the capacity to self-handle moisture.
   - The mat can be made of straw bales, wood chips, peat, or a high-humus and soil mixture that has good percolation.
   - Dig area to a 40 cm (16 in) depth below grade to accept the materials that will provide the initial sponge.
   - On hard or impervious ground, either plumb to a septic or sewer system, or dig a leachate absorption trench to accept excess moisture and divert it to an infiltration trench or pit system.

4. **Buffer Zone:**
   - There should be a fenced-in area surrounding the compost bins that limits access to the area most likely to be potential sources of pathogens.
   - Plan for a minimum of 1.5 m (5 ft) clearance around the sides and back and 1.75 m (6 ft) at the front.
   - Incorporate more space if moldering bins or commode buckets are to be stored for processing.
   - Use signage that notifies others that human waste is being housed or composted.

5. **Volume and Number of Composters (compost piles):**
   - Bins should be sized to accept the expected volume of materials over time, allowing enough room for human excreta, household compost (kitchen scraps), bulking agents, and covering materials.
   - Build a minimum of two, but preferably three, compost bins at 2 m (l) × 2 m (w) × 1.5 m (h) (72" w × 72" l × 60" h) to allow composting materials to fully mature over time (1–2 years).

6. **Dedicated Tools:**
   - Tools should be dedicated for use only for managing the composting — to limit the potential spread of pathogens.
   - Clearly label all dedicated tools.
   - Tools to dedicate include a hose with nozzle, a rake, and a pitchfork.
   - If required for transporting stabilized and unsanitized materials, a wheelbarrow must be dedicated for use only with the composting.

7. **Proper Setbacks:**
   Setbacks are specified distances that structures must be built away from some place deemed in need of protection; this could be a street, a river or other water body, a shore, or a flood plain. You will need to check your local jurisdiction to learn where you can legally put your compost pile. Where we live (in Canada) we have to locate our compost piles with the following setbacks:
   - 30 m (100 ft) from a source of drinking water or drinking water supply well.
   - 30 m (100 ft) from fresh water, seasonal fresh water, or sea water.
   - 15 m (50 ft) from a breakout (area where water openly seeps out of the ground).
   - 3 m (10 ft) from a property line.

Sealed Systems

There are options for completely sealed designs; some are manufactured (e.g. Earth Tubs), and there are many DIY systems out there. Figure 3.23 shows two such systems. One uses IBC totes (Intermediate Bulk

Hinged hatch
- ¾" marine plywood
- Sealed with exterior grade sealer

Vent
- 4" screened
- Vent on each side

IBC tote sealed compost bins
- 1000 L modified IBC totes
- Vented
- Lockable hatches
- Drained
- Can be cladded with wood or metal siding (not shown)
- Reduces cover materials required

Fig. 3.23: *Sealed compost bin system using IBC totes, corrugated culvert, or corrugated roofing; allow for drainage, locked latches, and excellent rodent control.*

ILLUSTRATION CREDIT: GORD BAIRD

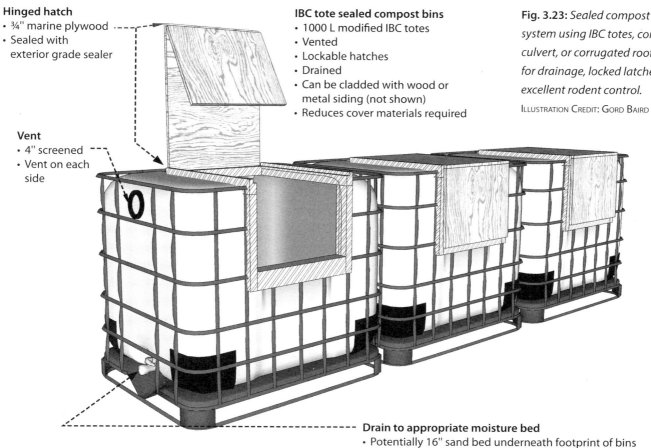

Drain to appropriate moisture bed
- Potentially 16" sand bed underneath footprint of bins
- Or leachate handling system

Corrugated culvert compost bin
- 4' or 5' in diameter
- Cutting procedure:
  1. Cut down the side of the hinges
  2. Weld (or attach) hinges
  3. Finish cutting the bottom and other side of the door

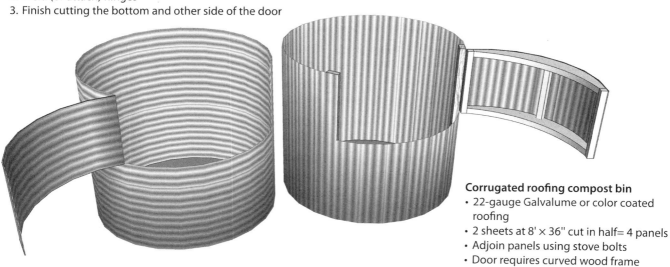

Corrugated roofing compost bin
- 22-gauge Galvalume or color coated roofing
- 2 sheets at 8' × 36" cut in half= 4 panels
- Adjoin panels using stove bolts
- Door requires curved wood frame

Container, or pallet container); the other is made from corrugated roofing materials.

Sealed systems require less cover material and they can be heated. It is relatively easy to drain leachate from the bin to an infiltration bed directly below the footprint of the bins; otherwise, a sealed system can be placed on impervious surfaces and drained to an appropriate leachate system. Bins can be cladded to meet one's aesthetic preferences, and they can be insulated to promote better heat retention. The downside is that removing the fully composted materials is more cumbersome due to smaller access panels.

Discharge of Solids

The choice of one of three options for final discharge into the environment will depend on the degree to which materials have been processed — that is, the degree to which we are certain that pathogen death has occurred or not, and how certain we are that the materials have or haven't been re-contaminated.

Offsite

Offsite discharge is simply having a certified contractor come in with their septic vacuum truck and sucking up the partially and incompletely processed materials. Considerations of design include ensuring easy and hygienic access for that collection. This would most likely be used with large continuous systems with chambers that promote the re-contamination of composted materials from the leachate fluids migrating into the finished product. Systems located in dense urban areas where there is not a feasible place to further process materials or apply it to the landscape would also be candidates for offsite removal.

Sometimes a highly engineered continuous system fails or requires servicing, and sometimes they need to moved or removed; therefore, planning to accommodate vacuum-truck access to the processor is a good idea.

On-Site Burial

On-site burial application is the burial of fully composted, sanitized, but untested materials. Consideration must be given to the burial site to ensure that it does not become *unburied,* i.e. it is not placed in an area prone to flooding, and it is far enough away from streams and drinking water sources (30 m [100 ft] is commonly given in the various regulations as the required distance). Burial should be between 15–30 cm (6–12 in) deep, in a trench rather than just a big hole (Ralston, 2016, p. 39).

This includes materials that have been processed to a matured and cured stage (as set out in Figure 2.13) and sanitized through one of the pathways discussed. Unless it is tested, never assume it is safe to apply above ground.

On-Site Land Application

On-site land application, or *surface discharge,* is the application of fully composted, sanitized materials that have met the conditions in Figure 2.17 or have been tested (proven safe) and meet the conditions in Figure 2.18 (Category AA standard) and Figure 2.19. These are basic requirements. There may be other environmental regulations in your area that disqualify *any* on-site surface application of *any* bio-solid, no matter what. Before you discharge, make sure to research the relevant regulations.

Photo of Heron Hall, by McLennan Design, a Living Building project, and their foam-flush toilet, gravity drop conveyance into a Phoenix continuous composter.

PHOTO CREDIT: DECLAN MCLENNAN

Photo of Heron Hall, by McLennan Design, a Living Building project, and their foam-flush toilet that feeds a Phoenix continuous composter.
PHOTO CREDIT:
DECLAN MCLENNAN

The Eco-Sense home, the first Petal-recognized Living Building home, relies on a batch commode Humanure system. PHOTO CREDIT: GORD BAIRD

Above: *An elegant humanure system at the Leckie residence, Saltspring Island, BC.*
PHOTO CREDIT: SANDRA LECKIE

Left: *Ann Baird's commode batch compost toilet prior to meeting Gord.*
PHOTO CREDIT: ANN BAIRD

Right: *The Desert Rain House, a certified Living Building, uses a Jets vacuum flush toilet.*
PHOTO CREDIT: TOM ELLIOTT

Top: *The authors, Ann and Gord with their dog Nina, sit in front of their solar-powered cob home surrounded by gardens.*

Bottom: *The authors' front yard has gone from a barren and very exposed landscape to a beautiful, abundant food-producing ecosystem in less than eight years.*

The McCaugherty Residence by Raincoast Homes is a showcase home relying on compost toilets.
PHOTO CREDIT: BILL MCCAUGHERTY

The George Residence uses compost toilets, rainwater, and greywater in their home on Gabriola Island, BC.
PHOTO CREDIT: LINDA GEORGE

The Bullitt Center — North America's most ambitious green commercial building.

PHOTO CREDIT: BENJAMIN BENSCHNEID

The Bullitt Center located in Seattle, Washington, a fully certified Living Building and considered the greenest commercial building in the world, relies on a Nepon Foam-Flush toilet hooked to a Phoenix processor.

PHOTO CREDIT: NIC LEHOUX

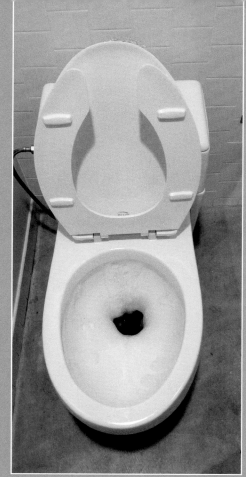

Left: *The Brock Environmental Center, a certified Living Building project, relies on a dry toilet connected to a Clivus Multrum.*
Photo Credit: Chris Gorri/ Chesapeake Bay

Right: *The inside of a Neptune foam-flush toilet.*
Photo credit: Glenn Nelson, Advanced Composting Systems Foundation/cbf.org

Below: *An Envirolet Flushsmart Vacuum Flush installed by Chris Magwood*
Photo Credit: Chris Magwood/ Endeavour Centre

A newly made IBC tote composter before being put into service.
PHOTO CREDIT: GORD BAIRD

Dumping large bins off a truck into an IBC tote composter.
PHOTO CREDIT: ANN BAIRD

Chapter 4

Design Considerations

Compost Toilet Selection Tool

This is likely the part of the book you will return to repeatedly as you consider your system. The objective of this chapter is to present ways to assess what systems and components will work for you personally and for your location.

So, what are the factors that influence one's decision to incorporate a compost toilet or not? The *Compost Toilet Selection Tool* given in Figure 4.1 incorporates philosophical goals, costs, site considerations, existing infrastructure, available power, climate, water, and other factors that will help you define and refine your choice of system. It is important to note that the various compost toilet types can be adapted in *many* ways — for such things as urine separation, type of toilet seat, amount of water, electricity and bulking agent requirements, addition of worms, and integration with existing systems you already have. Because there are so many possible combinations of systems and elements, you may wish to go through this simple tool a few times as your knowledge increases.

As mentioned in Chapter 1, we have included just a few names of particular toilets, meant as examples only. Many countries and geographic locations have different names for the same system. This book is meant to focus on compost toilet styles and methods, and not on specific brands.

With regard to comparative costs, we give price ranges for the various styles in Chapters 5, 6, and 7. The Selection Tool (Figure 4.1, see page 56 and 57) does not include price ranges, just cost comparison.

When looking at costs, it's critical to understand that there is a direct relationship between the amount of hands-on labor a system will require (servicing time) and the system's cost. In general, the more labor required, the cheaper the system will be.

Decision Factors
Philosophical
Resource Recovery

Many people who are interested in installing composting toilets are motivated by the desire to recover resources that would otherwise go back into our environment unused and in forms that cause harm. For this group of people, philosophy may guide them to the answer of the first Big Question for a CT installer: *Do you wish to recapture the resources contained in urine or feces — or both?*

If collecting urine for use as a fertilizer is the primary desire, there are two basic system options: urine diversion via a source-separating seat, or a dedicated separate urinal.

Whichever option you choose, you will need a urine collection and storage system and a urine reuse strategy that accounts for storing, diluting, pumping, and application. Urine diversion is discussed in greater detail in Chapter 8.

With urine out of the way, the options for feces are simplified. There are three methods to choose from:

- Batch commode
- Moldering
- Continuous

Compost Toilet Selection Tool		
Number		Chapter Reference
1		
1a	I want an inexpensive, simple method.	Go to #2
1b	I am ok with a moderate cost.	Go to #6
1c	I am ok with a more expensive system (least amount of ongoing work).	Go to #9
1d	I want all the benefits of a compost toilet but without any composting.	I want a high-tech toilet. Nano Membrane toilet (protype phase).	10
1e	I don't want to do any calculations, and I don't want a leachate system.	Go to #3 or choose Nature's Head.	7
2		
2a	I don't mind dumping fresh buckets (with urine and feces) into compost pile.	Go to #3 Commode Batch (Humanure).	5
2b	I want to separate my urine and only collect feces in a compostable bag (neatly placed in compost pile).	Go to #5 then go to #7
2c	I want to separate my urine and only collect feces in a small moldering system.	Go to #8
3		
3a	I want to buy a ready-made system.	Lovable Loo	5
3b	I want to custom-build a bathroom cabinet.	Go to #4
4		
4a	Custom cabinet bench with hidden toilet seat.	Eco-Sense Hidden Bench	5
4b	Custom cabinet in bathroom.	Eco-Sense Easy Access Cabinet	5
4c	Custom cabinet with large bins for higher use or less frequent servicing.	Commode Batch (Humanure) high volume	5
5		
5a	I want a ready-made manufactured system.	Separett or ECOJOHN	5
5b	I want to custom-build my cabinet.	Go to #4
6		
6a	I want an all-in, self-contained system.	Envirolet, BioLet, and Sun-Mar	7
6b	I want a simple urine-separating toilet (collect urine).	Nature's Head	7
6c	I want a simple urine-diverting toilet (urine to drain system).	Separett - also see #7	7
6d	I want a centralized and more conventional toilet.	See #9
7		
7a	I want to collect my urine.	See chapter 3 for details on urine collection	3
7b	I want to divert my urine.	See chapter 8 for details on diverting to existing sewer, greywater, or leachate systems.	8
8		
8a	I want to use a regular 5 gallon bucket for feces only and then allow it to sit until stabilized.	Modify the Eco-Sense cabinet design for urine diversion seat. Commode batch with urine separation seat.	5
8b	I want a self-contained simple manufactured system.	Nature's Head	7
8c	I want a larger moldering system. I will spend more money.	See #9

Fig. 4.1: *This Compost Toilet Selection Tool is helpful in narrowing down choices and guides you to pertinent sections of this book.* Credit: Ann Baird

Compost Toilet Selection Tool			
Number			**Chapter Reference**
9			
9a	I don't mind moving larger containers around and building a large cabinet under the toilet seat. I would like option of using worms.	See #10	
9b	I want a manufactured centralized system similar to my flush toilet. I want the option of using worms.	See #11	
10			
10a	Unsealed bins in sealed vented cabinet — simplier.	Custom-built sealed cabinet with movable large bins.	6
10b	I want sealed vented bins — more complex.	Custom-built cabinet with large, mobile sealed bins.	6
11			
11a	I want a manufactured carousel type.	Four options: Ekolet, Aquatron, EcoTech, Biorealis.	6
11b	I want a vault type.	Two options: Clivus and Phoenix.	7
11c	I want a simple mechanical system with regular toilet seat that moves solids into another composting place, and diverts urine to a leachate system.	Ecodomeo	6
11d	I want to flush.	See #12	
12			
12a	I want micro-flush on same floor as processor.	Jet vacuum flush. See options 11a or 11b	3
12b	I want micro-flush on second floor.	Micro-flush or foam-flush. See options for 11a or 11b	3
12c	I want to keep my conventional full flush toilet.	Aquatron	6

If the primary goal is collecting and composting *all* excreta (urine and feces) for full resource recovery, a thermophilic compost pile is your best bet, however you still have four options:

- Batch commode for everything (all-in)
- Batch chamber/moldering
- Vermicomposting
- Continuous (but with a separate composting process to meet surface discharge requirements)

Simplicity and Resiliency

For some people, the attraction of a composting toilet lies in its ability to simplify one's living environment (no sewer hook-up to worry about or septic system to install).

For some, the notion of improved resiliency is the attraction (if electricity goes out, there is no way to pump water to a regular toilet!). If your primary goal is simplicity and resiliency — no moving parts and limited technology — there are two basic options (which are also the least expensive):

- Commode batch with compost bins (e.g. Humanure bucket; Separett)
- Simple moldering systems (e.g. Humanure with urine diversion and larger receptacles; Nature's Head; Air Head; Biorealis)

Early adopter

Early adopters tend to have philosophical viewpoints that leads them to embrace and demonstrate innovative technologies before they are popular — and they are often willing to pay the higher price that comes with this. When it comes to CTs, they often want a system that resembles a conventional toilet — making the lever or push-button system a natural choice. There are *many* of these toilets on the market. Here a few types, installed in the world's most prestigious and advanced buildings (those that are certified Living Buildings), and in some of the world-class parks we ourselves have visited.

- Larger whole-house continuous systems (e.g. Phoenix, Clivus)
- Vacuum or foam-flush pedestal/conveyance (e.g. Jets, Wostman EcoVac, Neptune, Nepon, and other "ultra low flow" and "high-efficiency toilets," using less than 0.5L per flush)
- Mechanical systems (e.g. Ecodomeo)
- Advanced moldering systems (manufactured carousel-types like Aquatron, Ekolet, EcoTech)

Site

The location of the home or office in which you want to install a compost toilet will undoubtedly dictate some of your choices. Urban and rural locations each have a distinct set of factors to consider.

Urban Sites

Due to the easy access of sewer, water, and power, choosing to use a compost toilet in an urban area is often guided by ideology and/or the type of building.

Resilience — Local impacts from storms, earthquakes, or water rationing may lead some people to consider having a compost toilet in an urban area to ensure that even when something bad happens, they will still have safe and hygienic bathroom facilities.

Secondary Detached Suites and Tiny Homes — Many urban areas are starting to allow smaller in-fill housing on city lots. These are often referred to as *garden suites, granny flats,* or *tiny homes.* In some cases, it may be appropriate to consider a compost toilet rather than making new connections to city sewers.

Limited-Use Toilet — Other urban applications may include installation in structures that are minimally used (and that can be managed discreetly) to serve the intermittent needs in an outbuilding, for example, a compost toilet in a garden shed, workshop, or studio.

Toilets that fit within the urban environment:

- Batch commode
- Urine collection — small batch
- Small-volume compost pile
- Small bin moldering system
- Carousel toilet systems
- Large, centralized system (often the choice of early adopters)

Zoning — The small properties typically found in urban settings equate to closer neighbors, which often results in tight zoning setbacks from property lines. These smaller usable spaces can limit your choice of compost toilet system to those that do not require an external compost processing pile. Some zoning can be quite limiting — such as not allowing any kind of a backyard compost due to potentially higher rodent populations and potential odors. In these cases, extra design attention is required for the compost area. Some systems, though, lend themselves to use in highly zoned areas: *carousel systems; long-cycle moldering systems;* and *large, centralized continuous systems that have batch moldering to provide final treatment.*

Property Resale Values — When it comes to selling a home that has a composting toilet, it is perhaps highly likely that a new owner won't share the same ideology or enthusiasm of the owner who did the installation. However, in cases where metered usage is documented to produce large water and sewer fees for a home, the savings offered by a compost toilet could actually be a selling point. All compost toilet systems use less water and reduce flows to sewer systems. However, there are (at least) two types of compost toilet systems that can be good choices for the urban setting because they look more conventional and function similar to a regular toilet (thus, less likely to have a negative impact on property values): *vacuum and foam-flush*; and *centralized continuous systems designed to accept flush*. These are discussed in more detail in Chapter 7.

Rural Sites

Rural properties have the least constraints and the most options. Criteria a rural property owner might need to consider include access to bulking materials, power, water, soil, site layout, local regulations, and the big one, human involvement in the system. These are covered below.

Servicing issues will vary depending on what the definition of "remote" is. Cabins or homes that are boat-access only, or even hike-in cabins will often dictate a toilet that uses minimal bulking agents (or a local source), can have variable access to power and water, and is readily transportable or can be built on-site. Conventional septic systems may be out of the question in these remote locations. Remote locations often have little-to-no regulatory oversight for private systems or they require pre-engineered systems that they can be installed quickly and simply. Remote systems may also not receive regular maintenance, and, for lack of a better description, they tend to receive a lot of "shit and abuse."

One choice for remote parks or trails is the Ecodomeo Dry Toilet (a batch system). For remote cabins, options include commode batch (Chapter 5), moldering, Ecodomeo, carousel (Chapter 6), Nature's Head, self-contained continuous systems (Chapter 7), and bag and bury systems.

Power

In our opinion, all systems need fans and ventilation, so at least a minimal power source is required. Likely, all regulators will want to see fans as part of the system as well. However, that said, the USDA does have a Sweet Smell Toilet Installation Guide (SSTIG), detailing how to incorporate passive ventilation systems in which natural air currents are created to draw air out of a toilet box and out of the bathroom. Example details in the SSTIG include exposing a vent pipe to warmer temperatures (through sun or heated upper room) at the top of the pipe, thus creating a positive air flow, vent height variables, and understanding prevailing winds (for more on this, search USDA Sweet Smell Toilet Installation Guide www.nps.gov/public_health/info/eh/waste/sst%20installation%20guide.pdf or parkandrestroomstructures.com/docs/sstinstallguide.pdf).

Public utility power

Homes with access to public utilities allow for more power-intensive options tied to fans, heating elements, pumps, motors. This does not constrain systems, but it does allow the inclusion of add-on options that a manufacturer may offer, such as automated systems (motors); jet vacuum pumps, and heating elements.

Renewable energy systems

Off-grid and renewable energy systems in either AC or DC form require an energy budget to work within — conservation of electricity becomes a determining factor in system choice. These systems *do not require* heating elements to function (though many could potentially incorporate an element).

- Commode batch
- Carousel
- Simple moldering
- Mechanical (e.g. Ecodomeo)
- Vermicomposting
- Vacuum (Vacuum on Demand, VOC) with DC pump using the Jet collection system
- Continuous systems with manual hand cranks and gravity-fed drains

Water

By their very nature, CTs are very frugal with water use. That said, water is still required in varying degrees for different functions; even the so-called dry toilets that do not require water to flush will require water somewhere within the operation and management of the system. The following is a list of water usage in different systems.

At the Pedestal

- Commode batch systems don't require water for flushing unless there is a urine diversion seat.
- Pedestals with chutes will require weekly cleaning. About 2 liters (½ gallon) of rinse water is needed for cleaning the chute and the brush used to scrub the chute.
- Pedestals with micro-flush or vacuum systems will use between 0.2–1.0 liter

(1–4 cups) per flush, and, due to the nature of the bowl, perhaps ½ liter (2 cups) per day to scrub and rinse to remove smears.

Rinse Water at the Compost Pile

- Commode batch requires rinsing buckets and tools: about 1 liter (¼ gallon) per bucket.
- Moldering bins will need to be rinsed upon dumping into a compost pile (or after burial).
- Tools (rakes, shovels, pitch forks) will need rinsing.

Rinse for Urine Separation

One cup (250ml) — some manufacturers suggest this rinse after each urination, and others suggest once each day.

NOTE: There is some concern about collecting rinse waters and urine if storage for sanitization is part of the design. Fluid management will be discussed in Chapter 8.

Flush Water

- Vacuum flush use between 0.2 L to 1.0 L (1–4 cups) per flush
- Micro-flush
- Bidet (rinse water instead of toilet paper)

Systems that Use the Least Water

- Commode batch
- Moldering — all types
- Carousel direct drop
- Mechanical removal (Ecodomeo)

Flood Zone / Riparian Areas

No residence, industrial structure, or waste management/recycling system should be built within a 1-in-200-year flood zone. If your home has a basement and you are in a flood zone (or have had floods in the past), do not install any toilet system or storage and collection system in the area of flooding. Self-contained systems must be located on floors that are not prone to flooding, and compost processors (outdoor compost piles) need to be on elevations that don't flood. Set compost piles back 30 m (100 ft) from streams, ponds, rivers, wetlands, and well heads. There should be a setback of 15 m (50 ft) from any *breakthrough* (area where moisture appears to be seeping out of the landscape).

Soil

Soils appropriate for CTs need to have the characteristics that allow them to handle leachate or other effluent in an ecological manner.

Percolation test procedure

Percolation tests provide data on the ability of soils to accept and deal with fluids. "Appendix A: Percolation Test Procedure," provides the procedure for determining your percolation rate. Low percolation will either require amending the site (building soil) to allow it to handle fluids, or avoiding systems that have leachate flows.

Bedrock

Bedrock does not accommodate leachate fields, so any system with leachate should not be used on bedrock unless a special infiltration system is installed. If you have no soil and are on bedrock, you may wish to build your compost pile with a roof; this limits run-off from the compost pile. But you should probably also make plans to build soil. You might also consider an *incinerator toilet* (briefly mentioned in Chapter 6).

Ambient Temperatures

Temperatures will influence all aspects of system design. Climate hardiness zones can be used as a rough guide to thinking about expected winter freezing conditions, seasonal distribution of rainfall, and seasonal temperatures. Cold-climate composting is inherently different from composting in temperate climates.

Climate Hardiness Zones 1–5

Periods of weeks to months below freezing will impede regular cleaning and rinsing outside. Compost piles can freeze solid in the coldest of temperatures, making it more challenging to add batch-composted materials to a pile — but it is still possible if you design a larger compost collection area, giving you enough space to pile materials. You can expect longer compost times, in such a case. In below-freezing temperatures, there is no decomposing going on. In spring, the pile comes back to life.

If you live in a cold climate, you should avoid commode batch systems because they require frequent emptying. If your system design requires urine to be collected, avoid storing it outside; plan for it to be stored indoors until temperatures are higher than 5°C (41°F).

Good system choices for cold-climate composting would be: *continuous* (with chambers housed inside buildings); *moldering* (with collection bins housed inside); or systems that employ *mechanical removal* (Ecodomeo).

Sizing

This section is geared toward the *designers* of systems; it is not meant as a prescriptive "must do." If you are hired to aid a homeowner in choosing or designing a system, you will find these formulas useful in removing the guess work from your design. If a homeowner or a client is making a sizeable purchase for a special order of bins or a centralized compost chamber, there should be an hour of design time set aside to confirm exactly what is needed, including the sizes and number of bins, and the servicing frequencies relating to storage and processing. If you are a home-owner looking at simpler systems, this may be a section you wish to quickly scan. (However, it's an excellent opportunity to involve your school-age children or teenagers in demon-strating why it's important to study math.)

There is nothing worse than sitting down to do your business and finding that some-thing is horribly wrong — or horribly full. It's not the end of the world, but it can sure start your day off badly. A system needs to be able to safely and hygienically collect all that is deposited by the peak number of users for the bathroom, and it needs to account for the solids, any bulking additives, urine, and any flush waters that may be part of the compost toilet design. Figure 4.2 is a table showing the usage and volumes that would be expected for a residence. (The data are de-rived from averages of the North American diet. Yes, diet affects the volume of excreta.) the table can be used as a baseline to work from for designing facilities — with the

understanding that usage patterns/volumes for non-residential toilets are different. For example, public toilets at remote trailheads tend to collect less urine and more solids, as they are generally used in the morning (many folks are likely to pee in the bush later in the day); a public toilet at a day-use park tends to collect more urine and less solids.

Volume of Pee and Poo

The per person daily volumes of feces and urine production, and the corresponding amount of carbon that is added to the system with each are listed in Figure 4.2.

Calculations

This section can be read to get the basic idea for how to properly design or to assess a system. We need to know a few basic facts, like how much do we pee, how much do we poo, and, if we use bulking agents, how much is required. With this information, we can calculate a whole host of other useful data for the design. The system you ultimately select, whether you build it yourself or purchase an off-the-shelf system, will determine how much of the following you will need to understand. At this point, the goal is to un-derstand the basic concepts. Consider this section a reference tool to be used when and if you need it.

This section lists some useful equations; in Chapters 5, 6, and 7 we will use them to figure out container (bin) sizing, number of bins required, urine storage tank size, and the expected frequency of servicing the sys-tems. Possible variables are listed in the table found in Figure 4.3. These equations can also be used to assess and verify a manufactured system's claims (in essence, providing the same service as NSF certification — for a lot less money).

Fig. 4.2: *Residential excreta amounts.* Credit: Adapted from Ralston, Manual of Composting Toilets and Greywater Practice, 2016

Parameter	Feces and Toilet Paper, Per Day Per Person	Urine Per Day Per Person
	Adopted Value	Adopted Value
Wet Mass (g)	180 g (6 oz)	1500 g (53 oz)
Dry Mass (g)	50 g (1.8 oz)	60 g (2 oz)
Volume (L)	0.18 L (¾ cup)	1.5 L (6.3 cups)

Carbon Addition	Fecal Event	Urination Event
Average # of event/day	2	4
Volume of added carbon per event (L)	0.5 L (2 cups)	0.25 L (1 cup)
Total volume of added carbon, all events (L)	2 X 0.5 L = 1.0 L 2 X 2 cups = 4 cups	4 X 0.25 L = 1.0 L 4 X 1 cup = 4 cups
Total volume of ALL excreta collections (urine, feces, TP, & carbon)	(0.18 + 1.0) + (1.5 + 1.0) = **3.68 L/day/person** (¾ + 4) + (6.3 + 4) = **15.5 cups/day/person**	

NOTE: Drainage system size is not the same thing as *fluid storage sizing*; that is a different conversation, found in Chapter 8, Fluid Management.

To perform these calculations, you need to collect the data values for the items shown in the table. Following are brief notes for each item that appears in the table.

A = Additives/bulking agents

Additives, or bulking agents, are expressed as volume in liters (gallons).

In Figure 4.2, there is a listing of the additives per event (pooping or peeing), with an average expected volume per person per day. Following the examples from above:

• If sizing for a commode batch with everything in, we account for total additives

of 2.0 liters (0.528 gallons) per day. This includes 1.0 liter of bulking agent for total poo events and 1.0 liter for total pee events.

• If sizing for moldering systems where urine is diverted, we account for 1.0 liter of bulking agent per day.

• If we are accounting for urine storage ... STOP... we don't put bulking agents in urinals!

CC# = Number of collection containers

The number of collection containers is represented as a quantity.

Batching systems are reliant on having enough collection containers to collect and store the materials in the raw state until the containers are ready to be emptied (after which, they can be put back into the

		Description	Units	Commode Batch	Moldering	Continuous	Urine (Collection)
A		Additive volume/bulking agent $A = A_s + A_u$ A_s = Additive for solids A_u = Additives for urine	L/person/day gal/person/day	✓	✓	✓	✓
CC#		Number of containers (required over the period for full processing to occur)	# of bins	✓	✓		✓
CSs		Container Size for solids	L (gal)	✓	✓	✓	
CSu		Container Size for urine	L (gal)				✓
E		Excreta volume per person/day	L/person/day gal/person/day	✓	✓	✓	
H		Headspace required	L (gal)		✓		✓
O		Occupancy — # of people/users per day	# of ppl.	✓	✓	✓	✓
Ris		Rinse volume for solids (if required)	L (gal)			✓	
Riu		Rinse volume for urine (if required)	L (gal)		✓	✓	✓
Sf		Safety Factor % (can be used instead of headspace) e.g. 10% expressed as 110%	100% + Sf%		✓	✓	✓
ST		Stabilization Time	days		✓	✓	✓
T		Time Target for container replacement	days	✓	✓	✓	✓
U		Urine volume per person/day	L/person/day gal/person/day	✓		✓	✓
Vr		Volume reduction % expected over the Time Target (T)	1–Vr%		✓	✓	✓

Fig 4.3: *Required Data Collection Table: Gives the abbreviations used for the information required or to be solved for when designing systems.* CREDIT: GORD BAIRD

collection rotation). Obviously, commode batch containers can be quickly reintroduced, as all the processing is performed in a compost pile on a weekly cycle. Moldering toilets on the other hand require a period of time of storage for stabilization/maturation/sanitization to occur. To determine the number of containers required to collect and store all the materials for a full cycle, you need to know the number of days required for the stage of processing sought (the Target Time [T]), and you need to account for at least one additional container (the one actively being filled).

The following equation, with the most basic example, clearly demonstrates the "extra" bin.

Number of bins (collection containers), or CC#

= (# of days required for stabilization ÷ # of days to fill a bin) + 1

= (Stabilization Time ÷ Target Time) + 1

If it takes 1 year to fill a bin (365 days) and it takes 1 year (365 days) to stabilize, then you would have the following:

= (365 ÷ 365) + 1

= 2 bins — the one curing and the one being filled

This may seem silly, but the math will be important once we get into other examples.

CSs = Container Size (solids) and CSu = Container Size (fluids)

Container size is represented as volume in liters (gallons). Whether we are solving for container sizing for urine storage or solid storage, it is the same process. If we already know the container size we are using, then it allows us to solve for other unknowns like Time Target for servicing or the number of containers required.

E = Excreta volume per person per day

Excreta volume is expressed as liters (gallons). Pulling from the table in Figure 4.2, we can figure out the excreta volume per person per day. So, if we are sizing for a commode batch, and not diverting any urine, we would combine the values for the liquids and solids — for a total of 1.68 liters (0.444 gallons). If, on the other hand, we had a moldering system that just collected solids, we would calculate for 0.18 liters (0.048 gallons).

H = Headspace and Sf = Safety factor

Head space (H) is the height of space from the top of the container that is to be left empty. This volume of space is presented in liters or gallons. Designs need to account for allowing enough headspace to comfortably service any collection system. If using a commode batch system, there must be enough room to wipe our parts and have clearance from the contents. A moldering system needs room for air exchange under the lid, and continuous systems need room for the raking, turning, and mixing functions to occur. Urine and leachate storage is not any different.

The *Safety factor (Sf)* is expressed as a percentage of collection volume, rather than as a measure of storage capacity, like headspace. In other words, safety factor is another way of looking at that wiggle room you want to have, by overestimating the amount of materials one would collect, thus providing a buffer for the unexpected.

The safety factor does not necessarily apply to all systems. Let's use the example of a home that is designed for six users, but is hosting a wedding of 60 people. For the commode batch systems, this is almost a moot point; the safety factor lies in having several spare buckets on hand that can be swapped as needed. For a continuous system however, safety would be in the form of having additional head space built in to handle the peak load. Generally, the safety factor is most commonly applied to the large central systems.

We express safety factors as a percentage of collected volume, applied as a percentage above and beyond the volume of materials to be collected. If you want a safety margin of 10%, then you *overestimate* by adding an additional 10% of what you expect to collect: 100% of your produced volumes, plus an additional 10%, so you design for 110%. Example equations throughout the following chapters will demonstrate this.

O = Occupancy: Number of users

Toilet systems need to be designed for potential peak usage, NOT for present usage patterns. To explain this concept: a new home with four bedrooms may only have two people planning on living in it, but in the future the home might need to service many more people. We need to design for that potential. For this reason, regulators have formulas within their own standard practices, guidelines, or codes that they use to determine *potential future occupancy of a building*. Suffice it to say, we will not dive into those individual specific formulas — other than noting that occupancy is determined by both the number of bedrooms and the floor area of a building.

For the purposes of this book, and as a rule of thumb, you can roughly determine occupancy as two people for the first bedroom, and 1.3 persons for each additional bedroom, thus allowing a safety factor to be built into the design. (Design will be determined by the number of expected "events" per day, rather than by day. For residences,

you can figure on six events per person per day, as listed in Figure 4.2; but for public-use sites, a designer would study the type of usage pattern and apply 0.18 L feces and 0.25 L urine to determine how to size for usage. (For detailed discussion and design for public-use sites, refer to BC Government's, "Manual of Composting Toilet and Greywater Practice," section C-1.4.5, p. 57.)

Ris = Rinse volume (solids) and Riu = Rinse volume (urine)

Rinse? I thought compost toilets were designed to avoid the use of water! Generally, they are, but as compost toilets become more accepted, manufacturers are moving in the direction of offering more models or options that allow for the micro-flush of solids and the rinse of urine. Even source-separating urine-diversion seats often require a cup of water to rinse the urine collection portion after peeing. There are good reasons for this that are discussed under the section for urine diversion in Chapter 8. Rinse volume is not applicable to all systems, specifically those that have robust drainage. Primarily, it's the systems with urine storage that need to account for and accommodate rinse volumes.

ST = Stabilization time

Stabilization time (ST) is represented in days.

For systems that have a degree of processing occurring in the collection container, there will either be a recommendation from the manufacturer as to the period that is required before the stabilized materials are to be removed, or a set time frame based on the science of decomposition rates. (*NOTE:* If you want to store materials till matured or sanitized — which requires more time — then you can use that time frame as the "stabilization time.")

Recommendations for stabilization time:

- Minimum of 12 months (365 days) if the temperature of the stored materials stays above 10°C (50°F) during this time.
- Minimum of 18 months (540 days) if the temperature of the stored materials drops below 5°C (41°F) for more than 7 days in a row.

Refer to Figure 2.13 (maturation) and 2.17 (sanitization) for further detailed time options.

(*NOTE:* Remember that *stabilized* materials are decomposed, but not *composted, matured,* or *sanitized.* Another stage of processing is required, which generally involves a dedicated compost pile or placement into another container for more storage time.)

T = Time Target for container (bin) servicing

Target times (T) are represented in days. How long does it take to fill a container?

Batching chamber systems are like moldering toilets systems because materials are collected and held over time in some manner, which results in a servicing *cycle*; this cycle is expressed as time in days. Obviously, the servicing cycle and container sizing are directly related. You can either determine your required container sizing by knowing what the desired time target for servicing is, or you can determine your servicing schedule by knowing the container size.

Knowing your *container size (CS)* and your *time target (T)* allows you to determine the number of storage containers required over the period of time to hold all collections — from raw to the desired stage (stabilized, matured/cured, sanitized).

U = Urine volume per person per day (U)

Following the exact format as for excreta above, but if we are only designing urine storage, we would use *U.* The average urine produced per person per day is 1.5 liters (0.396 gallons).

Vr = Volume reduction factor

We have all seen compost piles shrink in size as the materials are changed. This volume reduction is expressed as a percentage of total volume of collected materials. There are two components to this: the factor describing how much a volume will decrease from the initial total; and the final volume of the decomposed material. In other words, if a pile reduces in volume by 40%, then at the end of the time period, the pile would be 60% of the original volume.

In a continuous system, decomposition is part of a continual process; therefore, we need to account for how much decomposition occurs inside that collection chamber *over a year.* If, over a period of a year, there is a 40% reduction in volume (a factor that is consistent over time) then a chamber can be sized smaller than it would if there was no volume reduction. Manufacturers will have figured this out when they recommend a system, but for those people designing their own system, *Vr* is a very useful calculation.

Volume reduction is expressed as the total initial volume collected (which is 100%), less the reduction in volume (e.g. 40%) equating to an actual final volume:

Volume total (%) − Volume reduction (Vr%) = Volume final (Vf%)

Example: 100% − 40% = 60%

(Where Excreta + Additive is represented as 100% and Vr is 40%)

Volume final (Vf) is $1 - Vr\% = 60\%$ of original volume

$Vf\% = 1 - Vr\%$

$Vf\% + Vr\% = 1$

System Sizing Data Collection Table

The table in Figure 4.3 should help guide you through which data you need to collect for the system you are designing for. Once you have this data collected, you would use the following calculations to solve for the unknowns in the table.

Calculations

Initial inputs

O = Occupants

Solved for either through the standard formula for the regulation you fall under or:

1st bedroom = 2 persons; each additional bedroom add 1.3 persons

$= 2 + (1.3 \, [\# \, \text{bedrooms} - 1])$

E = Excreta liters per person per day (or gallons)

E = value in Figure 4.2

U = Urine liters per person per day (or gallons)

U = value in Figure 4.2

A = Additives in liters per person per day (or gallons)

$A = A_{solids} + A_{urine}$ values found in Figure 4.2

If the system collects just solids (has urine diversion) then $A = A_{solids}$

Ri = Rinse volumes in liters

$Ri = Ri_{solids} + Ri_{urine}$ values found in manufacturers' specification

Ri_{urine} if unspecified can be 0.25 liters per event (1 liter/day/person)

Sf = Safety factor: recommend 10% to 15%

ST = Days to stabilization

ST = 365 days if the collection chamber or storage unit is $\geq$ 10°C (50°F) and does not drop below 10°C (50°F) for more than seven consecutive days

ST = 540 days if the collection chamber or storage unit is $\leq$ 9°C (48°F) for seven or more consecutive days

ST = or the time chosen to meet guidelines in Figure 2.13 and Figure 2.17

H = Headspace: recommend a minimum of 10 cm–15 cm (4"–6")

H is a volume calculation

Volume = Length × Width × Height

Liters = $L_{cm} \times W_{cm} \times H_{cm}$

Gallons = $(L_{inches} \times W_{inches} \times H_{inches}) \, 0.00433$

Container sizing (CS)

Container Sizing: Moldering

If working with a Safety factor %:

$CSs = (O \, (A + E + Ri_{solids}) \times T \times Vf) \times (1 + Sf)$

If working with a Headspace volume:

$CSs = (O \, (A + E + Ri_{solids}) \times T \times Vf) + H$

Container Sizing: Continuous

$CSs = (O \, (A + E + Ri_{solids}) \times T \times Vf) \times (1 + Sf)$

CS = mfgs. recommended # of daily users

Container Sizing: Commode Batch

$CSs = (O \, (E + U + A) \times T) + H$

Container Sizing: Urine

If working with a Safety factor %:

$CSu = (O \, (U + Riu) \times T) \times (1 + Sf)$

If working with a Headspace volume:

$CSu = (O \, (U + Riu) \times T) + H$

Target Time (T)

Target Time: Moldering

$$T = \frac{CSs}{0\,(Vf)\,(1 + Sf)\,(A + E + Ri_{solids})}$$

$$T = \frac{CSs - H}{0\,(Vf)\,(A + E + Ri_{solids})}$$

Target Time: Continuous

$$T = \frac{CSs}{0\,(1 + Sf)\,(Vf)\,(A + E + Ri_{solids})}$$

$$T = \frac{CSs - H}{0\,(Vf)\,(A + E + Ri_{solids})}$$

Target Time: Batch

$$T = \frac{CSs - H}{0\,(A + E + U)}$$

Target Time: Urine

$$T = \frac{CSu}{0\,(1 + Sf)\,(U + Ri_{urine})}$$

$$T = \frac{CSu - H}{0\,(U + Ri_{urine})}$$

Number of collection containers (CC#)

$$CC\# = (ST\ days \div T\ day) + 1$$

Number of Occupants (O)

In the case where you know the Container Size and the Time Target, but do not know how many occupants the system can serve, the equations can be solved for Occupants (O). An example of this might occur if you want to verify whether your toilet system can handle having extended family move in with you, if you are planning a larger family, or if you want to verify a manufacturer's stated claim. Here's an example equation for this calculation for a carousel system:

$$0 = (CSs - H) \div (E + A)T$$

All these equations might seem daunting. But there is no need to fret. In the following chapters we will give plenty of examples, demonstrating how to use these equations in practice. You can use our example equations for your own design by simply plugging in your information in exchange for ours.

Chapter 5

Commode Batch Systems

COMMODE BATCH is the simplest of all the compost toilet systems. It consists of a toilet pedestal (the "commode") positioned to allow deposits to go directly into a bucket, bin, or other container. The bucket/bin is simply removed and its contents (the "batch") dumped into a designated outdoor compost pile. All stages of processing occur in the same outdoor pile. With this system, there is not much requirement for math (no safety factor percentage to contemplate), just a headspace to manage.

The commode batch is commonly referred to as the *Jenkins Bucket* or the *Humanure system* because it was championed by Joseph Jenkins, author of *The Humanure Handbook*. Whatever term is used, the system is basically just a cabinet with a toilet seat that has a container underneath. The container size varies with the design. The cabinet itself can be custom fitted to the space; most hold a sawdust bin and spare bucket(s). When one bucket is full, the cabinet is opened, the bucket is removed and swapped out with an empty one.

For commode batch systems, it's important that the bucket or bin not become so heavy as to be hard to move, causing strains and injury. Weights of no more than 11 kg (25 lbs) should be the target. A standard 20 liter (5 gal) pail is ideal. If larger buckets or rolling bins are used, a special design may be needed at the compost processing pile to accommodate dumping out the larger container. This could involve ramps or even mechanical aids to safely place the bins in position for dumping their contents.

Many people use a compostable bio-bag to line their bins and buckets. We do not. They do offer the advantage of needing limited rinsing of the containers when dumped, but we find they interfere with the manipulation of the compost pile. The reason for the interference is that bags do not break down quickly enough. When you next open the compost pile to add new material, the bags "get in the way" of your pitchfork or rake tines.

Once the materials have been placed in the compost pile, they are covered with dry material — often, straw is used. The biological processes that take the raw materials through the stages from raw to stabilized and then to matured/cured and sanitized, occur in one spot. There are many videos online that show this process.

Key Considerations for Commode Batch Systems

- They require regular interaction or "servicing" by the user — to swap out full buckets.
- Easy access to the outside and proximity to the compost pile is needed. Stairs or uneven ground should be avoided. Tripping while carrying a bucket is not good.
- A dedicated space for a compost pile is needed close to the home.
- The receptacle or bucket must be small enough to handle safely.
- Water access and a dedicated hose for rinsing buckets and tools needs to be provided near the compost pile (where there are freezing temperatures in winter, hoses will need to be drained when not in use).

- Regular servicing naturally solves for peak usage, so sizing is a non-issue. Heavier usage just increases how often buckets are swapped or dictates the need for more buckets.

- Having multiple bathrooms is as easy as building additional cabinets and venting them.

- These systems are generally the simplest and most cost effective of all CTs.

- These systems require large volumes of bulking materials — in the form of coarse sawdust or wood chips. A supply needs to be kept near the pedestal, and a bigger supply is needed for the compost pile outside.

- Kitchen scraps including meats, bones, egg shells ,and veggie materials are all acceptable and useful to support the thermophilic processing in the compost pile.

- Due to frequent servicing, insects hatching inside the buckets (that have high moisture content) is not an issue.

Commode batch systems are ideal for residential full-time use that is serviced by an owner; they are also good for public facilities that have daily servicing by a dedicated person. They have the advantages of being very low cost, a very simple technology, and easily retrofitted to fit an existing space. They also offer very flexible peak usage. Another perk: items that might fall into the buckets (like dentures or glasses) can be easily recovered and disinfected.

However, commode batch systems are not ideal in situations in which guests have to service the system or in areas where there is no access to carbon bulking agents.

Manufacturers and other resources for the systems described below can be found in Appendix B. Online search terms that can be used to seek further information and design ideas are *batch*, *humanure*, and *bucket toilet*.

Humanure Bucket Systems

The commode batch can be broken into two sub categories: the bucket system and the (much larger) bin system. The bucket systems are small systems, generally site built, becoming part of the cabinetry. See Figure 5.1 for an example. The toilet seat is built into a cabinet that houses (preferably) two collection receptacles (usually 20 liter [5 gallon] buckets) and a compartment to house a bulking agent (e.g. wood shavings/sawdust). Access is directly at the toilet pedestal; therefore, the compartments need to allow for easy placement and removal of the buckets. When 75% full, these buckets weigh about 11kg (25 lbs), which is easy enough to handle without causing strain.

Venting can be simple, comprised of a pipe with a fan exhausting air to outside; this creates low pressure in the cabinet, which makes air flow *in* from the surrounding room, through the toilet seat, across the buckets, and then outside. This airflow pattern ensures that no condensation forms on the underside of the seat, and it removes humidity. The vent is screened at its exit with ¼" hardware cloth to stop vermin access, and screening is situated for easy removal and cleaning as needed (every couple of years) because dust and lint build up on the screen.

The compost pile is ideally located near the building so buckets do not have to be carried long distances; the pile must be located in an area not prone to flood. Following common practices of setbacks across the all industries dealing with water, wastewater, and sensitive ecosystems, all composters collecting feces require a 30 m (100 ft) setback from water courses and wells, and we suggest a best practice of 3 m (10 ft) setback from property boundary. The composting area should have a minimum of two — but ideally three — composters that are at least 1.5 m³ (2 yd³) in size; they must be encircled by a fenced compound (for more details, refer to the "Compost Processor" section near the end of Chapter 3, including Figures 3.21, 3.22, and 3.23).

Urine can be collected together with feces (all-in), or it can be partly or fully diverted as desired. If urine is collected all-in with the feces, then bulking agents will be used to soak up the fluids as required (an adult's usual output is 2 liters/day). Bulking agents serve many purposes; they add texture and coarseness, promote water absorption, improve aesthetics and odor control, and provide additional carbon for the composting process. If a urine diversion system is employed, less bulking agent is required and there are longer periods between swapping the buckets.

Servicing

The steps listed here are common to most bucket systems.

- Place empty buckets in the cabinet.
- When first bucket is two-thirds full, swap it with the spare (empty) bucket beside it. Do not place a lid on the full bucket, as you want the moisture to continue to be evaporated and vented, but you could cap it with another 1 inch of bulking agent.
- When second bucket is two-thirds full, remove both buckets to composting area and place lids on them.
- Put two empty buckets into the cabinet. (You should have enough buckets or bins on hand to handle one weeks' worth of collection.)
- Top up the bulking agent (sawdust) bin as required.
- About once per week (maybe less often), open up the compost pile by excavating a

depression in the middle of the pile about 16 inches deep, then dump the buckets into it. (*NOTE:* You will always have a thermometer in your pile, so remove it first and set it aside before opening the pile.)

- Rinse each bucket into the pile. Put 1 L (4 cups) of sawdust in the bottom of each bucket, and close each with a lid. Place them in the dedicated storage area. They are now ready for use when required.
- Bring the excavated materials back on top of the new additions.
- Rinse tools into the pile and return them to their designated spot and then cover with straw to provide 5 inches of cover.
- Place your thermometer back into the pile.
- After one year, quit adding to the pile, start a new compost pile, and let the old pile sit for one full year (or more).

Fig. 5.1: *A commode batch compost toilet in a cabinet that houses two 20 L buckets and a shavings bin; it is serviced by a ventilation pipe covered in fabric and accessed via a hinged seat that lifts up.*
Photo credit: Ann Baird

- Every year, harvest the older pile that has sat for one year with no new additions, allowing this space to start a new pile in its place. We usually do this in spring, as this is when the soils have warmed and the soil biology is more active, and it's when we are in our gardens and orchards.

There are many videos online showing this simple process. To watch Gord go through the steps, you Google "Eco-Sense Living" and "Humanure."

Sizing

The beauty of the compost batch toilet is in its adaptability. Due to the regular servicing and the ease of it, peak usage design is not required. But you do need to have enough buckets on hand to get the job done. In general, you need a minimum of 2.5 buckets/person per week. So, for a residence of four people, you would need 10 buckets.

These systems are adaptable to highly variable usage — a system design for two people can easily service one week with 6 people or an evening with 60 people.

Sizing calculations can be done as laid out in Chapter 4, but in all reality, due to the flexibility and ease of adding additional buckets into circulation, mathematical calculations are not required.

Multiple bathrooms

As mentioned earlier, it is relatively easy to have multiple batch compost bathrooms in a home. Each bathroom would have a dedicated cabinet, and the number of buckets remains constant, meaning that a home with 4 people would still use 10 buckets a week — whether it has one or more bathrooms.

Keep in mind that a single-bucket cabinet can be used as an *ensuite* just for nighttime urine. If you don't install a fan in the system, use double the shavings and service about once a week. Alternatively, you can line the base of the bucket with 4 inches of wood ash and continue to add the regular amount of shavings after each use, and then you might only need to service every three weeks. Note that if condensation occurs on the underside of the toilet seat, the toilet seat will start to smell like urine; in that case, you can increase either the ash/shavings or the servicing frequency. (See Figure 5.2.)

Cleaning and maintenance

- Clean the toilet seat as you would with any other system.
- Sweep loose sawdust as needed.
- Dump and rinse buckets outside.
- Clean dust from fan assembly when it starts to get noisy — about every year or two.

Power requirements

Low — from 0.5W–5W for fan that can minimally move 12–20 cfm; such fans are usually 20 dBa or less.

Water requirements

- Low — only required when buckets are dumped and rinsed on a weekly basis.
- One bucket receives about 20 deposits to fill up to two-thirds full. Each bucket takes 1 to 1½ L to rinse. This equates to 75 ml per use (5 tablespoons). This compares very favorably to flush toilets.

 Advantages:

- Simple
- Highly effective
- Low tech — no chance of mechanical failure
- Power efficient

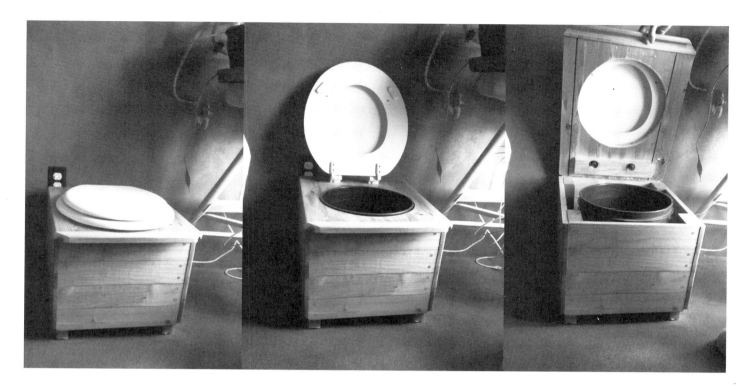

- Water efficient
- Excellent source for creating composted humus
- Odorless bathroom experience
- No sizing required and peak usage is not applicable
- Cost: $ ($500–$1,000 cabinet, buckets, venting [excludes expense of building composter])
- Minimal time commitment: A family of four will produce about 9–11 buckets a week; stored and dumped all at once, the process takes about 10–15 minutes/week.

 Disadvantages:

- Considerable amounts of bulking agents are required (about ¾ m³ [1 yd³] per person per year).
- Requires monitoring for bucket swaps.
- Requires routine weekly bucket dumping.

Fig. 5.2: *An ensuite commode used for nighttime collections of urine, using sawdust and/or ash, dumped weekly. No fan (electricity) needed.*

PHOTO CREDIT: GORD BAIRD

Designs

Following are three design options. The first is Joseph Jenkins's Loveable Loo; the other two are designs for different types of cabinets. (Note that resource information for any system mentioned can be found in Appendix B.)

Wherever costs are given, keep in mind that they do NOT include the cost of building the composter/compost bins themselves.

Loveable Loo

Cost: $ ($400 toilet, $200 venting and fan)

Joseph Jenkins sells the Loveable Loo, a commode batch kit, at the Humanure Store, along with accessories (compost thermometers, bio-bags, and books). This system can be used with or without compostable bags.

Eco-Sense Hidden Bench

Cost: $$ ($1,000 for cabinet materials, buckets, and venting)

Eco-Sense is our home and our business, and many of the designs we come up with for various items we label "Eco-Sense *X*". This design is one we use in our own home and have made for others. In this design, the toilet system is hidden within a bench; the upper lid (the bench seat) flips up, allowing access to the toilet and toilet paper. This is a useful concept for small spaces, especially in "tiny homes." This system can be used with or without compostable bags.

Design plans for this are found in Figure 5.3, Figure 5.4, Figure 5.5, and Figure 5.6.

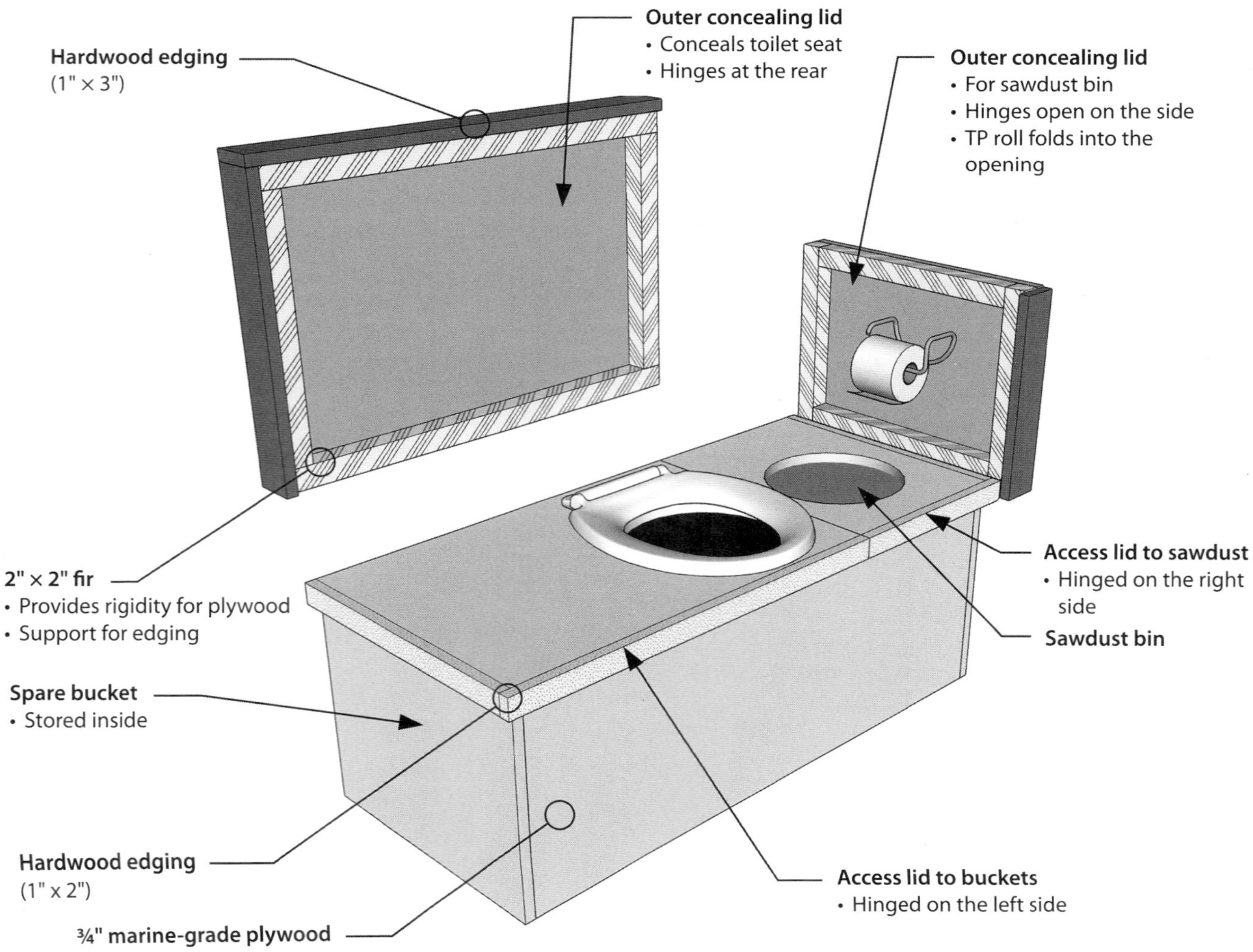

Fig. 5.3: *Eco-Sense design of a commode batch cabinet for two buckets and a sawdust bin.*

Illustration credit: Gord Baird

- **Hardwood edging** (1" × 3")
- **Outer concealing lid**
  - Conceals toilet seat
  - Hinges at the rear
- **Outer concealing lid**
  - For sawdust bin
  - Hinges open on the side
  - TP roll folds into the opening
- **2" × 2" fir**
  - Provides rigidity for plywood
  - Support for edging
- **Spare bucket**
  - Stored inside
- **Hardwood edging** (1" x 2")
- **¾" marine-grade plywood**
- **Access lid to sawdust**
  - Hinged on the right side
- **Sawdust bin**
- **Access lid to buckets**
  - Hinged on the left side

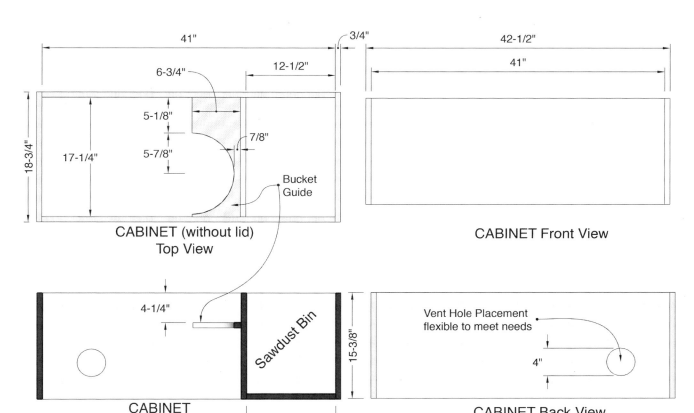

Fig. 5.4: *Eco-Sense commode design: top section.* ILLUSTRATION CREDIT: GORD BAIRD

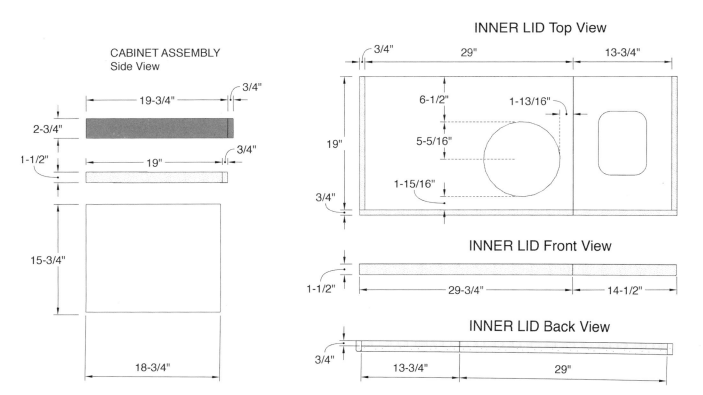

Fig. 5.5: *Eco-Sense commode design: concealing lid.* ILLUSTRATION CREDIT: GORD BAIRD

Notes:
- Plywood is ¾" marine-grade
- 2 × 2 SPF
- 1 × 2 Hardwood (edging)
- 1 × 3 Hardwood (edging)
- #8 2" wood screws
- Doweling
- Waterproof white glue
- Piano hinge (stainless or brass)

Inner lid(s) hinge on the left and right side of the cabinet.

Upper lid (large main lid) hinges off the rear of the Inner Lid.

Upper lid (small side lid) hinges off the right side of the smaller Inner Lid.

Additionally, a "hold fast" or "lid stay" can be used to fasten the large inner lid to the cabinet.

Locking pin — once completed, a locking pin can be installed to lock the large Upper Lid to the Inner Lid
(self-evident when completed).

Felt bumper cushions — installed on the underside of the Upper Lid. to provide protection to the lower lid.

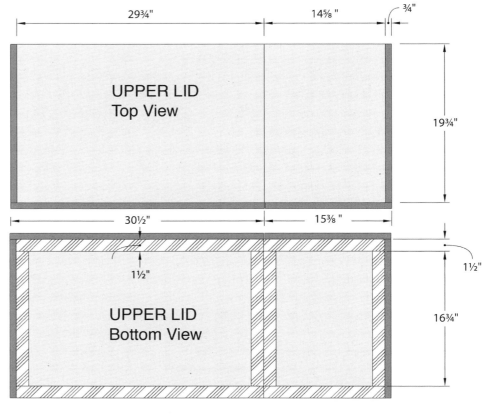

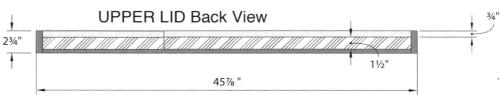

Fig. 5.6: *Eco-Sense commode design: front cross section.*

Illustration credit: Gord Baird

Eco-Sense Easy Access Cabinet

Cost: $$ ($1,000 for cabinet materials, buckets, and venting)

This cabinet is both elegant and sturdy; it is our preferred style at Eco-Sense, in part because it allows easy access for swapping buckets. This system can be used with or without compostable bags.

Design plans for this cabinet are found in Figure 5.7, Figure 5.8, Figure 5.9, and Figure 5.10.

Fig. 5.7 (right): *Eco-Sense Easy Access cabinet has a center opening lid, with center opening door, a bucket guide, and easy access to sawdust.* ILLUSTRATION CREDIT: GORD BAIRD

Fig. 5.8 (below): *Eco-Sense Easy Access cabinet front and top dimensions.* ILLUSTRATION CREDIT: GORD BAIRD

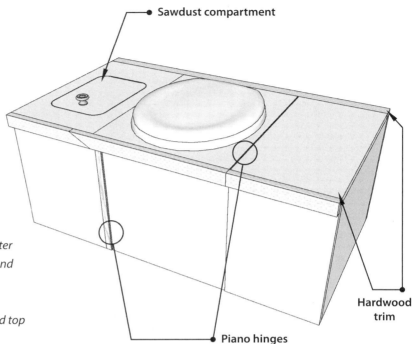

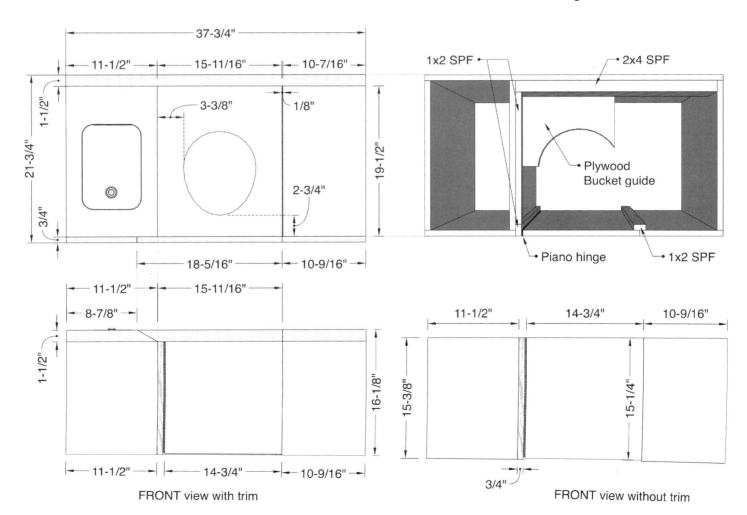

FRONT view with trim

FRONT view without trim

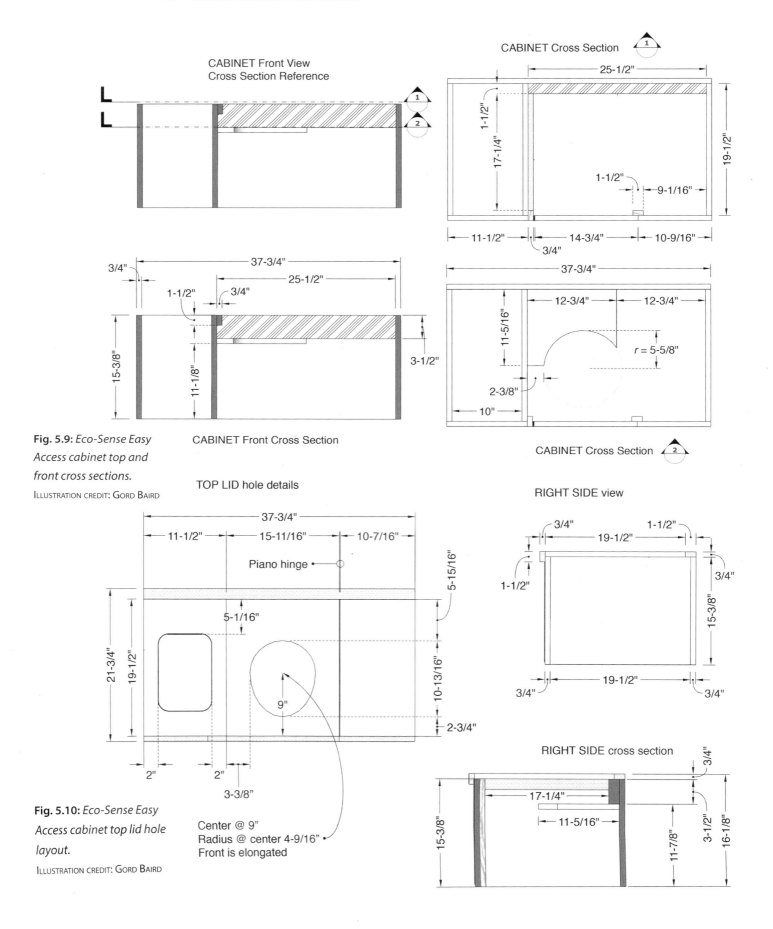

CABINET Front View
Cross Section Reference

CABINET Cross Section ▲1

CABINET Front Cross Section

Fig. 5.9: *Eco-Sense Easy Access cabinet top and front cross sections.*

ILLUSTRATION CREDIT: GORD BAIRD

CABINET Cross Section ▲2

TOP LID hole details

RIGHT SIDE view

Fig. 5.10: *Eco-Sense Easy Access cabinet top lid hole layout.*

ILLUSTRATION CREDIT: GORD BAIRD

Piano hinge

Center @ 9"
Radius @ center 4-9/16"
Front is elongated

RIGHT SIDE cross section

Manufactured Commode Batch Systems

Cost $$ ($1,200–$1,500)

Manufactured batch systems are very much the same as above: they have a pedestal and they collect into a container as a "batch," which is composted. What makes them different? They come with a toilet seat pedestal rather than a cabinet (Separett and ECOJOHN both have batch system models). Privacy flaps are usually built into the pedestal; when one sits, the pressure on the seat opens the flap, when you get up, the flap closes and hides the contents. A collection chamber collects solids into a compostable bag, while urine is diverted via a urine diversion toilet seat as seen in Figure 5.11 and Figure 5.12. This diversion of urine leads to less frequent servicing than the commode batch (Humanure) system because it avoids the storage of fluids and the added bulking agents needed to manage them. Some models do come with an added heating element under the waste box to aid in drying the materials further. Fully 75% of feces is water, so a heating element can play a big role in such a system. Plus, sometime urine does not always end up in the correct place, so a drying element can be doubly useful.

When the bag housed in the pedestal is full, it is moved to a composting pile, the pile is opened, bag and contents are deposited, then covered. Very simple. And no buckets to dump.

Collected urine can be drained or collected.

Drainage system options include:

- Diversion to an existing sewerage system
- Diversion to a greywater system
- Diversion to a dedicated leachate drainage system

Fig. 5.11: *Separett toilet collects materials into a bag in the bucket, has a privacy flap, and a urine-separating seat.* Photo credit: Gord Baird

Fig. 5.12: *Urine diverter tube: Some have a pressure-fit seal to the toilet for servicing.* Photo credit: Gord Baird

Collection options:

- Collected, stored, and used seasonally as a fertilizer
- Collected and used directly on the compost pile to aid composting
- Collected and dried (systems are being developed that use ash (not charcoal) to absorb the urine and heaters and a mixing system to aid in drying it all out.

Venting installations would follow the guidelines set out in Chapter 3. The manufactured systems come with their own fans, wherein duct work could tie into existing bathroom moisture fans or routed direct to the exterior of building.

Servicing

The steps listed here are common to most manufactured batch systems.

- Install a bag in the collection bin/box.
- Monitor the level of collections and remove bag when full.
- Deposit the full bag into a compost processing pile.
- Install a new bag.
- Clean inner seat as required.
- Rinse the urine collection with a cup of water after each urination.
- Replace any "chemical bio pucks" for the urine catchment trough as per the toilet manufacturer specifications. The chemicals in various "bio blocks" include bacteria and enzymes and chemicals to soften the water. Sadly, some contain fragrances. Always look to choose ones with no fragrance, perfume, scent, or phthalates; these are dangerous toxic chemicals, and they may kill good bacteria (ask to see the MSDS sheet [material safety data sheets]).

- Flush urine tube (seen in Figure 5.12) annually with an acid, such as strong cleaning vinegar and scrub the urine diversion tube with an appropriately sized bottle brush.

Sizing

- Adaptable to highly variable use — higher use leads to more frequent bag replacement.
- Urine collection requires design work — use the Container Sizing equation for urine (CSu) to calculate required storage tank size (calculations in Chapter 4).
- Urine collection also requires a sanitization plan before its end use (if not disposed of regularly into the compost pile).
- Emergency drainage into any existing sewerage, greywater, or leachate systems may be a wise consideration.
- Secondary urine storage is required if you need to store and sanitize it before use; this requires calculating the number of Collection Containers (CC#) (calculations given in Chapter 4).

Multiple bathrooms

Because these are *stand-alone* systems — each bathroom can have its own toilet unit. Batch systems are highly adaptable to all homes in any climate for any retrofit.

Cleaning

Urine diversion systems require frequent attention, including rinsing with water, removal and cleaning as required, and periodic cleaning of urine diversion tube.

Some urine-diverting seats have a replaceable curtain valve similar to those used in waterless urinals; these can require replacement yearly (a consumable product you buy, replace, and dispose of).

Power requirements

- Low to moderate.
- Fans ranging from 0.5W–5W for fan that can minimally move 12–20 cfm; such fans are usually 20 dBa or less.
- Heating elements are optional; can be AC or DC, 45W–55W.

 Advantages:

- These pedestals are highly adaptable, and can be adapted to work with other composting systems discussed in the following chapters.
- Extremely easy to retrofit into existing bathrooms.
- Suitable for cold climates (no storage of full buckets required while waiting for compost pile to thaw).
- With urine diversion, the fill time is drastically extended from a couple days to a couple weeks (as there is little to no bulking agents added).

- If bio-bags are used, then no water is required to rinse collection buckets.
- Cost is low relative to standard toilet: $$ ($1,200–$1,500).

 Disadvantages:

- Ongoing expenses for purchases of supplies like bags or urine diverter "chemical" pucks are a consideration.
- One must plan ahead to ensure there are always spare bags (something akin to realizing you're out of toilet paper in the house at the wrong time).
- More toilet seat cleaning and urine diversion system cleaning.
- Compost piles without the urine do not reach consistently high temperatures and may require longer periods (6–12 months) of additional composting before they should be considered sanitized.

Humanure Bin Systems: High-Volume Commode Batch

Cost: $$–$$$ ($1,500–$3,000)

The high-volume version of the commode batch system (Humanure) is exactly the same as the Humanure Bucket system except it utilizes larger bins, which decreases servicing intervals; and, due to size and placement of the bins, the access to them is from below, in a different room. This different room is in reality a small sealed cabinet that is just large enough to house the large bins — 84 liter bins (22 gallon) rather than 20 liter (5-gallon) buckets. The pedestal can be situated on the floor directly atop the cabinet room with the bins below, or the upper portion of the room/cabinet can extend up through the bathroom floor and become part of the

pedestal. If the cabinet becomes part of the pedestal, care has to be given to the design to keep the seat height at the standard level (18"). See Figure 5.13.

Bins need to be accessed; therefore, the room/cabinet needs to be designed for easy access. For this reason, the doors to the room/cabinet are often designed to be accessed from the exterior. If having access to the bins from the exterior of the building is part of the design, then consideration needs to be given to ensure its access doors are exterior-grade quality.

Because urine diversion can be accommodated, this type of batch system is easily modified to become a moldering system.

The area that houses the compost bins requires a design that allows for easier dumping of the larger bin. Having a working

Fig. 5.13:
Design elements for high-volume commode batch system using wheeled bins.

ILLUSTRATION CREDIT: GORD BAIRD

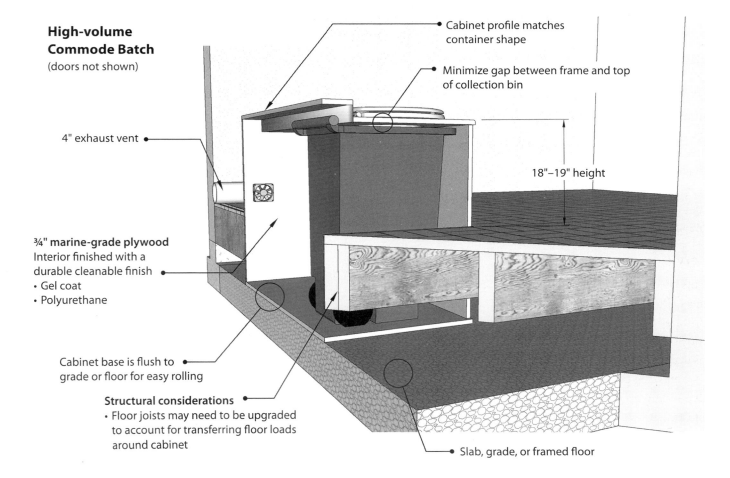

High-volume Commode Batch
(doors not shown)

Cabinet profile matches container shape

Minimize gap between frame and top of collection bin

4" exhaust vent

18"–19" height

¾" marine-grade plywood
Interior finished with a durable cleanable finish
• Gel coat
• Polyurethane

Cabinet base is flush to grade or floor for easy rolling

Structural considerations
• Floor joists may need to be upgraded to account for transferring floor loads around cabinet

Slab, grade, or framed floor

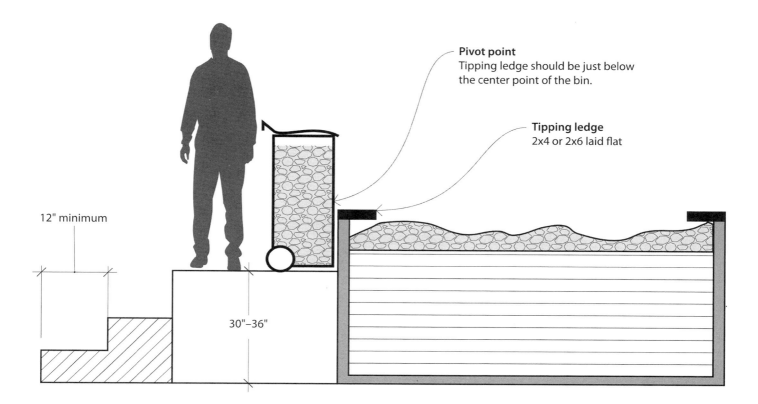

Pivot point
Tipping ledge should be just below the center point of the bin.

Tipping ledge
2x4 or 2x6 laid flat

12" minimum

30"–36"

platform elevated .75 m–1 m (30"–36") off the ground adjacent to the compost pile allows for easy and controlled tipping and rinsing. Access of the wheeled bin to the platform is best achieved via two to three steps with a tread depth of 12" each, rather than a ramp (see Figure 5.14). If a ramp is used, follow your Building Code's accessibility guidelines; the ramp should not exceed 6% slope.

Servicing

Bins are swapped when one reaches desired level (75% is a good target), and bins are dumped into compost pile. Why 75%? Any more volume than this makes handling the bins very difficult.

Servicing could be weekly to monthly depending on usage and urine diversion options.

At the compost pile, some method must be devised that allows tilting the bins into the pile rather than lifting them, as the weight is too much (the author places these rolling bins onto a pickup truck tailgate, backs up to the compost pile, and easily tips the bin's contents).

Sizing

Completely flexible, as with all batch systems.

Multiple bathrooms

- Not easy to retrofit.
- Does require special attention to carpentry and cabinetry.
- Retrofit requires modifying floor joists, adding header, doubling joists which can involve removing wiring, plumbing, and other fixtures for the modification.

Fig. 5.14: *Stepped platform for dumping wheeled bins.* ILLUSTRATION CREDIT: GORD BAIRD

Cleaning

- Regular cleaning of toilet seat, as for any toilet system.
- Upon removal of a ¾ full container, additional carbon can be added to create a 2-inch sawdust cap atop the contents to ensure there is no slosh or fluid splash.
- The rim of collection bin is sprayed with a cleaning vinegar (24%), household hydrogen peroxide, or chlorine bleach solution to keep crevices clean and remove opportunities for flies to breed.

Power requirements

- Fans — from 0.5W–5W for fan that can minimally move 12–20 cfm, is 20 dBa or less.
- Optional cabinet heating to help with evaporation.

Water requirements

- Likely the lowest water use of all systems — approximately 2 liters of water is required to rinse a bin when dumped.
- Averages less than 30 ml (2 tablespoons) per bathroom use.

 Advantages:

- Highly flexible system that allows opportunity for adjusting between processing styles (batch and moldering) (see Chapter 6).
- Infrequent servicing due to larger collection bin.
- Extremely efficient water use.
- Cost: $$–$$$ ($1,500–$3,000).

 Disadvantages:

- Bins are heavier to manipulate and require more thought with the design of composting pile.
- Longer storage times require close attention to moisture, as too much moisture can breed flies.

Chapter 6

Chambered/Moldering Batch Systems

A MOLDERING TOILET is a system that separates and diverts urine away from the solids and decomposes the solids in a chamber for an extended period of time at non-thermophilic (cool) temperatures. Without the addition of fluids and the corresponding additives associated with them, the containers fill much more slowly, over a period of weeks, months, or even years. At the end of a collection cycle, the collection chamber is removed from service and set aside for 12–18 months to stabilize, or 2–3 years to cure and *sanitize*. These designs require some form of ventilation to allow gas exchange and excess moisture to escape — all while keeping critters and the weather out.

After a period of 12–18 months of decomposition (moldering), the material is considered stabilized, but not necessarily cured or sanitized. (Refer to composting basics given in Chapter 2.) Materials can be tested to see if they are safe to introduce to the landscape. If they are not, they can be moved to a compost pile for quicker processing to a sanitized state, or they can be held longer in the container to sanitize. Once emptied, a chamber is ready to be put back into service.

The moldering concept can be applied to relatively small systems, such as those with moveable 80 L (21 gallon) containers, all the way up to large vaults that are stationary (in the latter case, it is the toilet pedestal that is swapped out [relocated]).

Systems incorporating swappable bins carry the same flexible serviceability associated with commode batch systems in that peak usage is not really an issue because you can swap out collection containers as needed and scale the system up by simply adding more collection containers. It provides the benefits of the continuous system by reducing the frequency of servicing.

A potential drawback — or benefit, depending on how you look at it — is dealing with urine collection. If you feel the need to splash into fluid management, Chapter 8 dives into the depths of it.

Moldering systems that incorporate built-in vaults have limited appeal for residential applications; hence, most people use bins for collection. Use of a dedicated vault requires more design forethought up front. Vaults require added precautions and detailing and the sealing off of the structure (building) from humidity and gasses. A minimum of two vaults is needed. Materials are directed to one of the two vaults either by way of a moveable pedestal, a diversion flap that deflects one way or another, or two pedestals; the one not being used is locked to keep it out of service. For these reasons, these systems are usually stand-alone systems used in public spaces, so they are not covered in this book.

Key Considerations for Chambered/Moldering Batch Systems

• Design needs to account for the time and temperatures required in a moldering plan. Refer to the "Non-thermophilic conditions" in Figure 2.13.

- Structures need to be designed around bin sizes, providing easy access directly under the toilet pedestal for bin placement, and allowing easy maneuvering of full bins to dedicated storage areas.
- For a sanitized outcome, design the moldering plan to the timeframes as set out under "Non-thermophilic conditions" in Figure 2.17.
- Materials can be transferred at any time to a compost pile; the system ceases to be "moldering," and is then treated like a commode batch, which is composted in thermophilic (hot) conditions.
- Methods to sanitize can be applied at any time during the curing process, as discussed in Chapter 2.
- Moisture management is a critical part of a moldering design (combination of urine diversion, leachate drains, fans, and perhaps even a heating element).
- Ventilation of aging containers requires alterations to lids to allow air exchange, prevent precipitation from entering, and control for pests.
- Bins stored outside should have their own fenced area (preferably with a cover), to protect bins from being accidentally knocked over.

Manufacturers' contact information and further resources can be found in Appendix B. Online search terms that can be used to seek further information are *moldering toilet, carousel toilet,* and *Full Circle Composting Toilet.*

Moldering systems are ideal for:

- Remote locations
- Homes with basement area directly below the toilet pedestal
- Cold climates (as well as warm ones)
- Extreme water conservation
- Pest and vermin control
- Small lots
- Recreational use
- Full-time use
- Retrofits — highly adaptable

Moldering systems are not ideal for:

- Public use (if urine diversion is part of the design, male misuse by standing rather than sitting results in excess moisture within the collection bin and the need for daily cleaning).
- Slab-on-grade buildings (no space under the toilet).
- Settings that require multiple toilets (most applications are designed for direct drop from the pedestal).

Moldering Design 1: Unsealed Bins in Sealed, Vented Cabinet

Cost: $$–$$$ ($1,500–$3,000)

In these systems, the bin(s) are situated directly under the toilet seat in a small cabinet designed to hold at least the active container, though preferably also an additional spare empty, as illustrated in Figure 6.1. The cabinet is screened and well sealed to avoid air leakage, and it is ventilated to let the air drawn in through the pedestal act as the primary air flow. One or two large cabinet doors provide access to the inside of the cabinet from the back. Additionally, this cabinet can also house the fan and any battery or electrical equipment required to run the system. If urine diversion is part of the design, the urine collection tank if small enough, can be situated in this cabinet. Practically, however, urine storage will be in a container larger than comfortably fits in the cabinet. If urine is collected into a small tank, and not used regularly, it will require transferring to a larger storage tank.

Access to the cabinet needs to be easy. For designs where there is a crawl space, the cabinets can be built to be either on grade or with a ramp at no more than a 6% slope to allow the collection containers to be

Fig. 6.1: *Moldering toilet vented cabinet with rear access, unsealed container, and urine storage collection.*

ILLUSTRATION CREDIT: GORD BAIRD

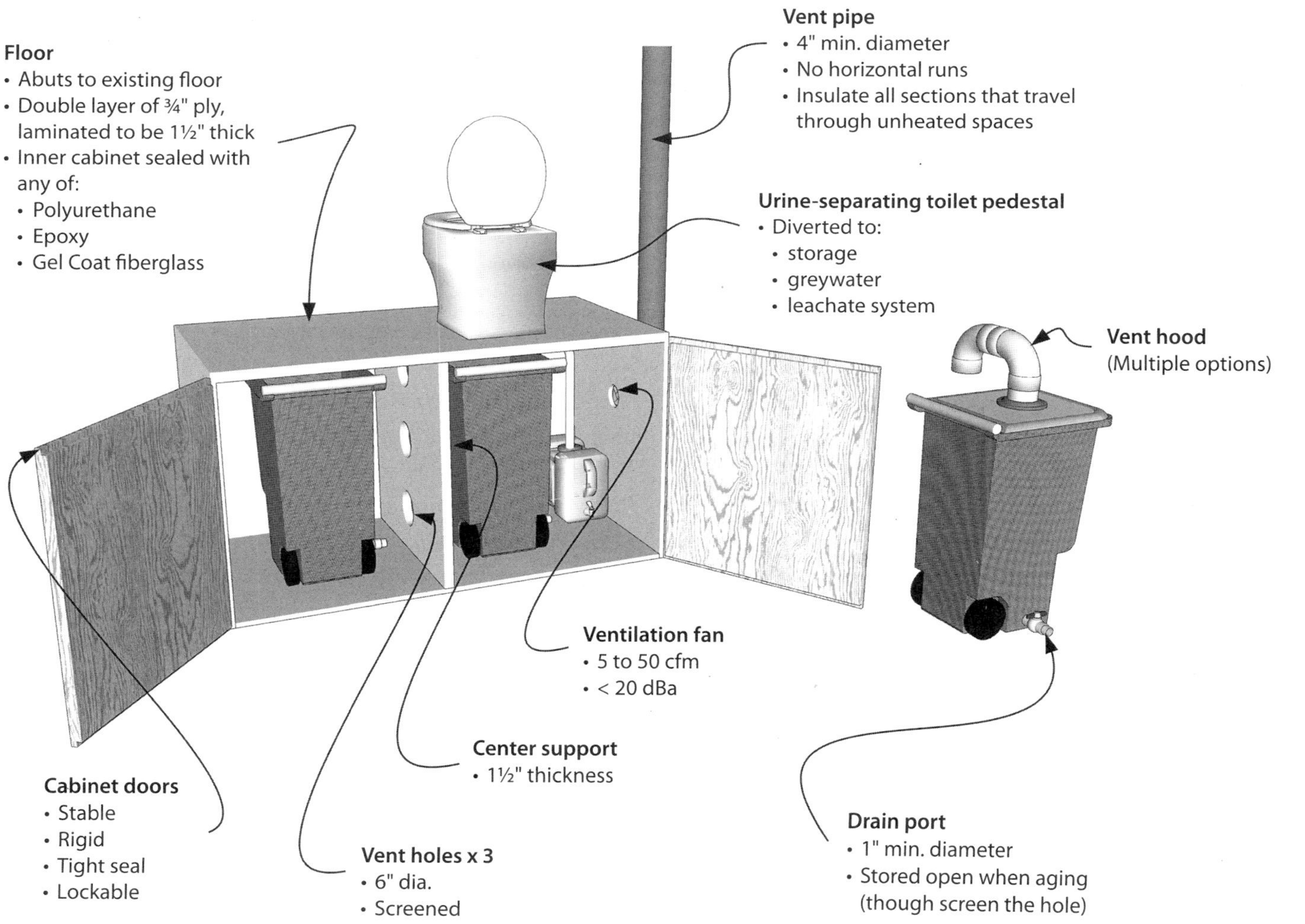

Floor
- Abuts to existing floor
- Double layer of ¾" ply, laminated to be 1½" thick
- Inner cabinet sealed with any of:
  - Polyurethane
  - Epoxy
  - Gel Coat fiberglass

Vent pipe
- 4" min. diameter
- No horizontal runs
- Insulate all sections that travel through unheated spaces

Urine-separating toilet pedestal
- Diverted to:
  - storage
  - greywater
  - leachate system

Vent hood
(Multiple options)

Cabinet doors
- Stable
- Rigid
- Tight seal
- Lockable

Vent holes x 3
- 6" dia.
- Screened

Center support
- 1½" thickness

Ventilation fan
- 5 to 50 cfm
- < 20 dBa

Drain port
- 1" min. diameter
- Stored open when aging (though screen the hole)

easily rolled out to the storage compound. If the building has a greater height under the pedestal (e.g. full-height basement) that does not allow a compact cabinet that snugly fits the bin, then there will need to be a chute from the pedestal to the bin very similar to the Full Circle design illustrated later in the chapter.

Storage bins set aside to molder can be vented in various ways, as seen in Figure 6.2. Option 1 can be a screened vent hood that attaches to the lid; option 2 would be a screened opening cut into the lid.

Moldering bins require a drain port near the bottom of the bin; the added benefit of a drain port is that once a bin is DRAINED and set aside for moldering, the port can be left open (if screened) to allow additional gas exchange (also illustrated in Figure 6.2). There should not be an issue of leakage from the drain port, but keeping the bins within a dedicated area ensures that a practical safety measure is taken.

The number of containers required is a function of the container size, the number of occupants, and the period of storage. Once

Fig. 6.2: *Venting options for wheelie bins used for moldering compost toilet systems.* ILLUSTRATION CREDIT: GORD BAIRD

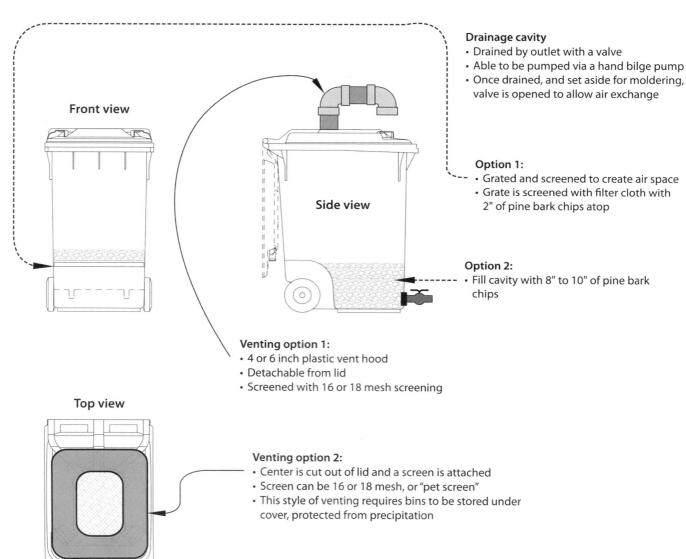

Drainage cavity
• Drained by outlet with a valve
• Able to be pumped via a hand bilge pump
• Once drained, and set aside for moldering, valve is opened to allow air exchange

Option 1:
• Grated and screened to create air space
• Grate is screened with filter cloth with 2" of pine bark chips atop

Option 2:
• Fill cavity with 8" to 10" of pine bark chips

Venting option 1:
• 4 or 6 inch plastic vent hood
• Detachable from lid
• Screened with 16 or 18 mesh screening

Venting option 2:
• Center is cut out of lid and a screen is attached
• Screen can be 16 or 18 mesh, or "pet screen"
• This style of venting requires bins to be stored under cover, protected from precipitation

Front view

Side view

Top view

a container is filled, it needs to sit for the specified period of time noted in Chapter 2 (Figure 2.13) to achieve the desired state of maturation.

Cabinet design

- Sized to house one or two receptacles of choice, as seen in Figure 6.3.
- Vented with a fan rated between 5 and 20 cfm, with a noise rating < 20 dBa.
- Easy access to leachate drain (30 cm/12" clearance from cabinet side wall).
- Floor that spans the cabinet (that the pedestal sits atop of) is able to handle normal loads, preferably 1½" thick, comprised of two layers of ¾" plywood laminated together with a center support underneath.

- Cabinet interior is sealed with a durable waterproof coating (polyurethane, gel coat fiberglass, or epoxy).
- Doors are durable, non-warping, locking, and create a tight seal (with weather strip).
- Doors are also weatherproof and insulated if they are on the exterior of the building.
- Access needs to be on grade or have a ramp that allows easy removal of full containers

Sizing: Solids collection containers

Choose your container size based on the following calculations. Smaller containers lead to more frequent swaps and more containers, but they are easier to move. The following example solves for bin size, but it could equally solve for the frequency of bin

Fig. 6.3: *Front cross section of moldering toilet vented cabinet for unsealed containers and urine storage collection.*

ILLUSTRATION CREDIT: GORD BAIRD

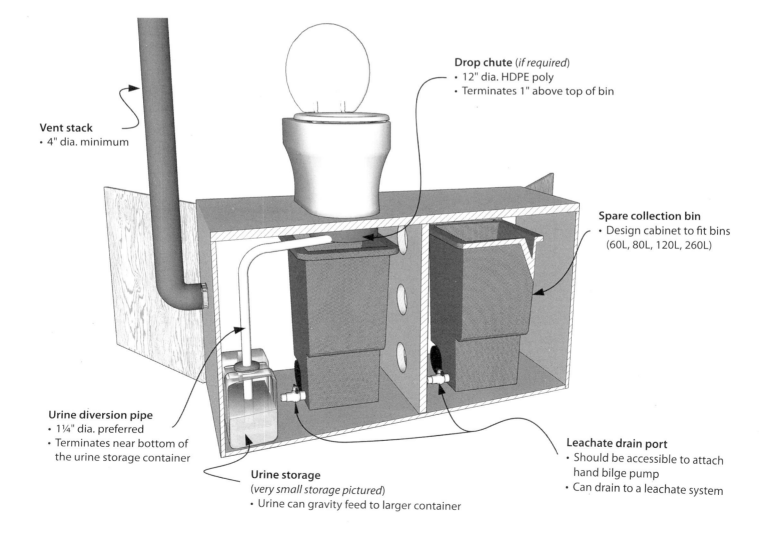

Drop chute (*if required*)
- 12" dia. HDPE poly
- Terminates 1" above top of bin

Vent stack
- 4" dia. minimum

Spare collection bin
- Design cabinet to fit bins (60L, 80L, 120L, 260L)

Urine diversion pipe
- 1¼" dia. preferred
- Terminates near bottom of the urine storage container

Urine storage
(*very small storage pictured*)
- Urine can gravity feed to larger container

Leachate drain port
- Should be accessible to attach hand bilge pump
- Can drain to a leachate system

servicing for a pre-chosen bin size. The table in Figure 4.3 will be a useful reference for the following examples.

Data for finding correct Container Size for the solids (CSs):

E = Excreta volume	0.18 L/person/day
A = Additive volume (minimal if any)	0.25 L/person/day
O = Number of occupants	3 people
Ri$_{solids}$ = Rinse volume/person/day	0 L
H = Headspace needed	(4") or 8 L
T = Time target for container replacement	60 days
Sf = Safety factor	Not required, as highly serviceable
Vr = Volume reduction %	Assume 0% for this example
Vf = Volume final is 1–Vr%	If Vr is 0 then Vf = 1

Calculation: *What Container Size for the solids (CSs) do I need?*

CSs = (O (E + A + Ri$_{solids}$) × T × (Vf)) + H

CSs = (3 (0.18 + 0.25 + 0) × (60) × (1)) + 8

CSs = 85.4 L

How many containers would be required for 12 months of collection and 12 months of storage? (At which point, the first filled bin can be put back into the rotation.)

Calculation: *How many Collection Containers (CC#) (or bins) are required?*
CC# = (Stabilization time in days ÷ Time Target to fill a bin) + 1 bin in use

CC# = (ST ÷ T) + 1

CC# = (365 days ÷ 60 days/bin) + 1 bin

CC# = 7 containers

U = Urine volume L/person/day	1.5 L/per/day
O = Number of occupants	
Ri = Rinse water L/person/day	0.25 L
Sf = % of container left unfilled as safety margin	10% (expressed as 110%)
CSu = Container Size urine	25 L

Urine collection

- Toilet pedestal has built-in urine diversion seat.
- Urine diversion pipe is a minimum of 1¼" in diameter, preferably 1½".
- The urine diversion pipe terminates just above the bottom of the tank, as seen in Figure 6.4.
- The seal between the urine collection tank and the pipe that enters it requires a very tight friction fit or other form of seal to maintain a fully sealed environment. To make it easy to service plumbing (e.g. moving a tank or swapping urine diversion piping to another tank), Banjo Cam Lock couplers or true union connectors can be installed just ahead of where the piping enters the tank.
- The urine diversion/collection system is vented (pressure equalized) as per concepts presented in Chapter 8 in Figures 8.8 and 8.9.

Sizing: Urine collection containers

Data needed to find correct Urine Collection Time Target (T) for servicing: (see table bottom left)

Calculation: *How many days (T) until the urine container needs to be swapped?*

CSu = (U + Ri)(O)(T)Sf

25 = (1.5 + 0.25)(3)(T)(100 + 10%)

T = 25 ÷ ((1.5 + 0.25)(3)(110%))

T = 4.3 days (less, if you drink a lot of beer)

Calculation: *How many collection containers are required to store urine for 90 days?* (CC#)

CC# = (90 days ÷ 4.3 days) + 1 container in use

CC# = 22

Using a small container like the one in this example (and the one shown in Figure 6.4)

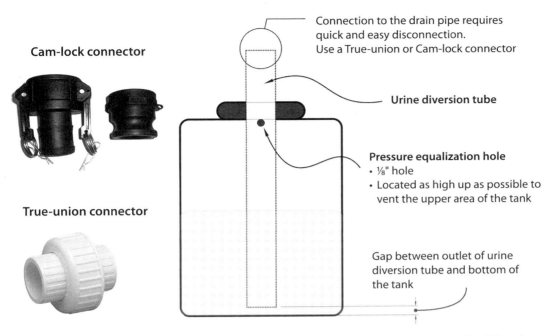

Cam-lock connector

True-union connector

Connection to the drain pipe requires quick and easy disconnection. Use a True-union or Cam-lock connector

Urine diversion tube

Pressure equalization hole
- ⅛" hole
- Located as high up as possible to vent the upper area of the tank

Gap between outlet of urine diversion tube and bottom of the tank

Fig. 6.4: *Cross section of a urine diversion storage tank with pressure equalization hole.*

Illustration credit: Gord Baird

leads to more frequent servicing as compared to larger systems. Obviously, space and preference will determine if frequent servicing is preferable or if larger collection containers (as shown in the next design) are more appropriate. Many aspects of these designs are interchangeable.

Leachate drainage

Leachate drainage is an optional way to *get rid* of excess fluids. The next section discussing the sealed moldering system, gives details on setting up a leachate system. It will cover options of gravity-fed drainage to a leachate field or leachate storage tank, and the use of a hand bilge pump to transfer leachate to one of those two systems. Sizing of leachate drainage to a small leachate seepage pit is covered in Chapter 8.

Servicing and cleaning

Servicing and cleaning will be minimized if:

- The urine diversion system is functioning well and if people's behavior patterns are correct (males sit to pee).

- An appropriately designed fluid/leachate collection cavity is present in the base of the solids container as illustrated in Figures 6.2 and 6.6.
- The drain port easily connects to DWV (drain-waste-vent) piping that drains to the leachate system, and is able to also easily connect a hand-operated diaphragm bilge pump to suck out minor blockages.

Successful servicing and cleaning comes down to being able to access all components *easily.*

Cleaning the leachate collection cavity in the bin — Once materials have aged and the bin is emptied, cleaning would involve dismantling the screen/grate system, hosing it off, and rinsing the bin completely; or, if the option of pine barks chips is chosen, it is as simple as installing a new layer of chips.

If a bilge pump is used, then a cycling of the pump through a sanitizing solution such as cleaning vinegar (24% acid), peroxide, or chlorine bleach solution after each use is required.

Insect control

If all aspects of the leachate system are properly connected, drained, and cleaned, insects will not be a problem. Some people take the step of weather-stripping the toilet seats to avoid vector access, but the resulting lack of air flow past the seat can lead to condensation building up on the underside of the toilet seat lid.

Power

Leachate systems do not require power other than that for a fan, unless a collection storage container is lower than the drainage to the leachate system. In this case, power would be needed for an automated effluent pump.

 Advantages:

- Simple management.
- Built from easily accessible common materials.
- Flexible — can be used as both a large commode batch or moldering system.
- Draining the excess fluids that build up in a bin allows the system to function better (keeps materials from being starved of oxygen, reduces odors, and reduces opportunities for insects to breed).
- Cost: $$–$$$ ($1,500–$3,000).

 Disadvantages:

- Added complexity if urine storage is incorporated (need to plan for storage, usage, and urine-diversion system maintenance).
- Requires a leachate system if there is no access to an existing sewer system.

Moldering Design 2: Sealed, Vented Bins

Cost: $$$–$$$$ ($4,000–$6,000)

The following design, illustrated in Figure 6.5, incorporates concepts from several different compost toilet systems (Vera Toga 2000, Ekolet carousel, drum composting). Perhaps one of the best thought-out moldering systems, it is comprised of a sealed vented bin and the conveyance method from the pedestal to the container is sealed to the bin; ventilation is integrated within the sealed system.

This design is more complicated than the first moldering design for two reasons: there are more modifications to the large collection containers — which range from 250 L (60 gallons) to 300 L (80 gal); and the ventilation ductwork is more involved. This said, the system is completely closed off to all vermin, is fully vented, and provides exceptional air circulation to the aging bins, making it an ideal choice for cold climates because moldering processes can occur within a heated indoor environment.

Full Circle Composting Toilets sells a manufactured version; and the Cape Cod Eco-Toilet Center offers information online on a hybrid version for systems and ventilation connections. Thanks to the Cape Cod Center and their open-source sharing, we are able to draw from their efforts and present this information (see Barnhart and Maingay, 2015).

Additional information on the Cape Cod Eco Toilet hybrid design and the Full Circle systems can be found at:

capecodecotoiletcenter.com/types-of-eco-toilets/hybrid_compost_toilet_systems/

fullcirclecompost.org/wp-content/uploads/2012/12/Full-Circle-Composting-Toilet-drawings-and-specifications.pdf

The special venting design that ties together ventilation and air recirculation offers complete control of air flow. The fan draws air down through the toilet and chute and cycles this air through the aging bins before being exhausted. This provides an ideal cold-climate moldering arrangement. Figures 6.5, 6.6, and 3.18 illustrate air flows, drainage, and container construction.

Toilet pedestal

A urine diversion toilet seat is required in moldering systems. Searches online will provide an array of companies that make versions that are ceramic, fiberglass, or plastic. A well-designed and easily available pedestal in North America is the Separett Villa 9220 (Figure 3.7) designed for below-floor collection systems. Ekolet and Wostman both supply ceramic pedestals to North American distributors.

Venting

Venting for this second moldering system relies on a central fan box that has the following features:

- The fan box has two sections: the lower suction chamber, and the upper exhaust/pressure chamber.
- The 4" vent stack exits out of the top chamber of the fan box and terminates outside.
- Entering the fan box from the bottom is a 3" suction pipe.
- The suction pipe collects airflow from the suction manifold that connects to all bins (the active one and those undergoing stabilization).
- The urine storage tank has a ¼" vent tube that also connects to the suction manifold, allowing urine tank pressures to remain stable (if urine is diverted to leachate, or

greywater, then urine storage is, of course, not required).

Venting for the active bin and the stabilizing bins are a bit different.

Active Bin

- A 3" flexible suction hose connects the bin to the suction manifold (attaches to the lid).
- The suction in the hose causes the air to be sucked into the toilet pedestal, through the drop chute, into the top of the active bin, across the contents, then into the suction manifold and through the fan box.

- The diagrams (Figures 6.5 and 6.6) show an additional 2" air port at the side; this is not used during the collection process (only during the aging process) — it stays capped when not in use.

Stabilizing Bin(s)

- The air flow pattern is reversed from that of the active bin — the air travels instead from the *top* of the maturing materials to the bottom, and *then* out. See Figure 6.5.
- The supply of make-up air (air that is required to replace the air that is sucked out) comes from the fan box upper chamber (exhaust/vent chamber).

Fig. 6.5: *Closed-container, vented moldering system: Combined design concept from Full Circle Composting Toilets and Cape Cod Eco-Toilet Center.*

ILLUSTRATION CREDIT: GORD BAIRD

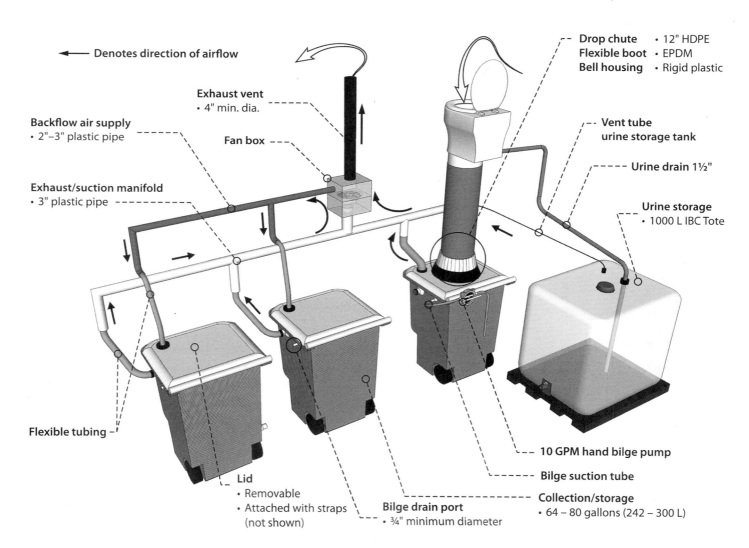

◄── **Denotes direction of airflow**

Drop chute • 12" HDPE
Flexible boot • EPDM
Bell housing • Rigid plastic

Exhaust vent
• 4" min. dia.

Backflow air supply
• 2"–3" plastic pipe

Fan box

Vent tube
urine storage tank

Exhaust/suction manifold
• 3" plastic pipe

Urine drain 1½"

Urine storage
• 1000 L IBC Tote

Flexible tubing

Lid
• Removable
• Attached with straps (not shown)

Bilge drain port
• ¾" minimum diameter

10 GPM hand bilge pump

Bilge suction tube

Collection/storage
• 64 – 80 gallons (242 – 300 L)

- This make-up air is recycled, drawn from the main vent stack.
- Suction of air from the bottom of the collection bin aids in evaporating any residual excess leachate that cannot be pumped out.
- The flexible suction hose is attached to the side suction port (which was capped off during collection).

One of the benefits of this system of sealed bins is that airflow to moldering materials does not require additional fresh, heated air; instead, oxygen-rich air is efficiently reused, which reduces the excessive heat loss that could occur from too much ventilation.

Containers

Readily available 64 gallon (262 L) wheeled trash receptacles or any heavy-duty large receptacle can be adapted to the above design principles. Wheeled garbage cans are available at retail hardware stores or from online container manufacturers and suppliers. Appropriate containers can be found by searching online for *bear-proof trash cans, tilt-trucks, wheeled truck,* or *wheelie bins.*

Container servicing time frame

If we know the container size, we can solve for the time frame it takes to fill it using the very same equation we used to calculate

Fig. 6.6: *Interior plumbing of leachate drain and air vent for a sealed moldering bin.*

ILLUSTRATION CREDIT: GORD BAIRD

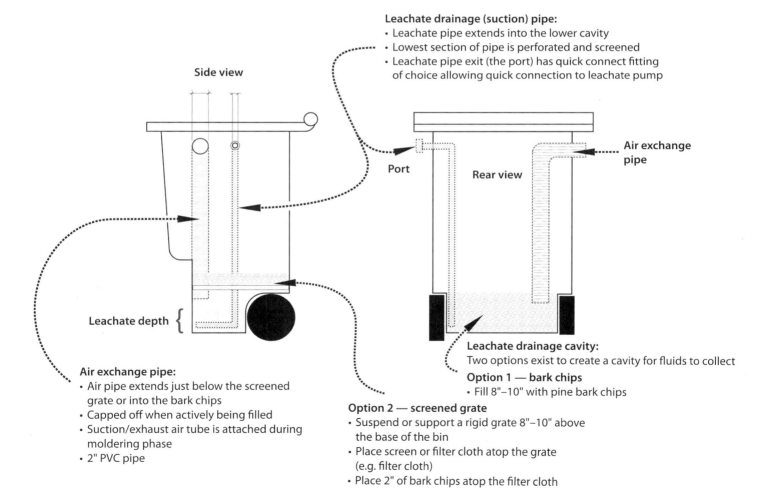

Leachate drainage (suction) pipe:
- Leachate pipe extends into the lower cavity
- Lowest section of pipe is perforated and screened
- Leachate pipe exit (the port) has quick connect fitting of choice allowing quick connection to leachate pump

Side view

Port

Rear view

Air exchange pipe

Leachate depth {

Air exchange pipe:
- Air pipe extends just below the screened grate or into the bark chips
- Capped off when actively being filled
- Suction/exhaust air tube is attached during moldering phase
- 2" PVC pipe

Leachate drainage cavity:
Two options exist to create a cavity for fluids to collect
Option 1 — bark chips
- Fill 8"–10" with pine bark chips

Option 2 — screened grate
- Suspend or support a rigid grate 8"–10" above the base of the bin
- Place screen or filter cloth atop the grate (e.g. filter cloth)
- Place 2" of bark chips atop the filter cloth

Container Size. In this example, we will solve for the Time Target (T), and will assume we are using a 64 gallon (262 L) receptacle as illustrated in Figure 6.6.

Data needed to answer: *How many days does it take to fill a container? Or: what is my Time Target?*

CS = Container size	262 L (64 gallon)
E = Excreta volume	0.18 L/person/day
A = Additive volume (if any req'd)	0.25 L/person/day
O = Number of occupants	3 people
H = Headspace needed	(4") or 40 L
T = Time target for container swap	Unknown
Sf = Safety factor	Not applicable
Vr = Volume reduction %	For this example we assume 0%
Vf = Volume final is 1–Vr%	If Vr = 0 then Vf = 1

Calculation: *Solving for Time Target for servicing interval (T):*

$CS_s = (E + A)(O)(T)(Vf) + H$

$262 = (0.18 + 0.25)(3)(1)(T) + 40$

$262 = 1.29T + 40$

$222 = 1.29T$

$T = 222 \div 1.29$

$T = 172$

Expect a bin to be filled every 172 days

Calculation: *How many Collection Containers are required?*

How many bins are required if we fill one every 172 days? The calculation for the number of containers required for 12 months of collection, when allowing 12 months of storage time (ST) to achieve maturation

(ST=365 days). (At this point, the first filled bin can be put back into the rotation.):

$CC\# = (ST \div T) + 1$ bin in use

$CC\# = (365$ day $\div 172$/bin$) + 1$ bin in use

$CC\# = 2.12$ bins $+ 1$

$CC\# = 3.12$ bins

Urine collection

Toilet pedestal has built-in urine separation seat.

- Urine diversion pipe in this design is 1½" in diameter (see Figure 6.5).
- The urine diversion pipe exits as near the bottom of the urine collection tank as possible (to maintain a sealed environment).
- The connection between urine collection tank to the pipe that enter through it has a very tight friction fit or other form of seal (Uniseal or threaded bulkhead fittings will serve this purpose).
- The urine collection container has a ¼" tube that connects to the suction manifold to provide minimal venting and pressure equalization.

How to size urine collection system components:

Data needed to find correct Container Size for the urine (CSu): (See table bottom left)

Calculation: *How long does it take to fill 1000 L container (T)? :*

$CS_u = (U + Riu)(O)(T)Sf$

1000 L $= (1.5 + 0.25)(3)(T)(110\%)$

1000 L $\div ((1.5 + 0.25)(3)(110\%)) = T$

$T = 173$ days (roughly 5¾ months)

It would take three people 173 days to fill the 1000 L storage tank shown in Figure 6.5.

Obviously, if one were to have an additional 1000 L tank, once the collection

U = Urine volume L/person/day	1.5 L/per/day
O = Number of occupants	3
Riu = Rinse water for urine L/person/day	0.25 L
Sf = % of container left unfilled as safety margin	10% (expressed as 110%)
CSu = Container Size urine	1000 L
T = Time target for container replacement day	?

container was full, it could be pumped into the other 1000 L storage, stored for 3 months (90 days), and used.

Leachate Pumping (optional)

This design includes a hand pump designed for boat bilges, rated at 10 gpm. Bilge pumps work well because they are designed for harsh environments and can handle particulates (suspended solids) — which describes leachate very well. Refer to Figure 6.6 for the following design description.

The leachate pump (not pictured) connects to the leachate drain port exiting the side of the container through some form of quick-connect fitting (cam-lock connector or true union). The leachate drainage (suction) pipe extends down to the bottom of the container and has a perforated and screened foot that sits horizontal at the base. The upper solid materials are blocked from entering this lower leachate collection via a 2-layer screening system consisting of a structural grate covered with a mesh or a deep layer of pine bark chips.

Structural grate — A plastic or fiberglass grate like those used in driveway drainage systems can be modified to fit in the base of the container, suspended off the bottom using spacers, such as 4" plastic pipe cut to the length desired for the height of gap required. If 6" is the desired height for the leachate collection area, then several pieces of pipe would be cut to 6", placed on end, and the grate would sit atop the cut pipe.

Screened mesh — Atop the grate is ⅛" screened mesh. This mesh can be HDPE plastic, stainless steel, or other durable non-corroding material. Avoid fragile mesh, like window screens.

The leachate pump (bilge pump) can be mounted on a moveable bracket that can hang from the bin being pumped, or permanently mounted to a wall.

If leachate is being pumped to a leachate system, then obviously it makes sense to have the wastewater pipe close by the pump.

Multiple Bathrooms

- Theoretically, these systems could be designed for multi-floor applications, but the reality is that the pedestal and chutes have to be placed directly over the bins, which can make a retrofit impractical or impossible.
- For public use facilities — where there are multiple toilets on the same floor level — this design could potentially be used.

Cleaning

- General hygienic cleaning, as with any toilet system.
- Cleaning of chutes and or flaps may require scrubbing with cleaning agents and a long-handled brush.
- Diverter seat requires frequent attention; it needs to be rinsed after each use, removed and fully cleaning as required, and the urine diversion tube requires periodic descaling.

Servicing

- If not hooked up directly to an automatic leachate drain system, easy and convenient access must be provided to hook up the pump and attach it to the leachate waste — ease of servicing for whomever is tasked with it lends to better sanitary procedure. The frequency of draining will depend on usage patterns; the best case scenarios requiring no pumping.
- Upon the filling of the collection bin, it is removed and replaced with a new bin.
- The newly filled bin is set up with a new lid assembly and connected to the ventilation system.

NOTE: If urine is going to be stored for longer than a of couple days, it needs to be stored for 3 months to ensure it self-sanitizes against pathogen growth that may occur once it leaves our body.

- Leachate is stable after aging minimum of:
  - 1 year if aged at ≥ 20°C
  - 18 months if aged at ≥ 5°C
  - 24 months if aged where temperature drops < 5°C for 7 days or more
- Leachate is sanitized:
  - If ≥ 5°C (≥ 41°F) temperatures are maintained, then 24 months since the last addition, OR
  - If < 5°C (< 41°F) temperatures are experienced, then 36 months since the last addition.
- Sanitization methods can be applied at any time during the curing process.
- Emptied bins are rinsed, grates and pipes cleaned, and readied for use.
- To avoid chlorine bleach, you can use a high-strength vinegar or Aseptox (a one-step peroxide powder), which can be found at beer- and wine- making supply stores.

Insect Control

There is very little need for insect control due to the well-sealed nature of the design.

Power

Power is needed for fans, but surprisingly little for such an intricate system: just a single 5-watt fan (12–20 cfm) will do the job.

 Advantages:

- Costs are lower than for a continuous systems (if self-built)
- Excellent ventilation and odor control
- Excellent insect control
- Great cold-climate option
- Worms can be added to speed decomposition
- Flexible usage patterns for high-volume applications

 Disadvantages:

- Intricate design
- Requires a broad range of skills and knowledge to piece together plumbing connectors, plastic molding (bell housing), and bin drainage system
- Cost: $$$–$$$$ ($4,000–$6,000)

Moldering Design: Carousel

Another form of moldering includes a rotating collection containment system that is, in essence, a large container with multiple compartments inside. These compartments are filled up sequentially; when one becomes full, the unit is rotated to move the adjacent empty compartment under the pedestal. The new, empty compartment is locked in position and readied for collection. Like all moldering systems, the full compartment is given time to stabilize and cure before it is emptied and ready to be refilled. Leachate drainage is built into the system. This is a totally enclosed system that provides exceptional sealing from insects and allows for excellent ventilation.

These types of systems have become one of the most widely installed in Scandinavia, with installs of one manufacturer's model alone numbering over 35,000 (no, that is not a decimal error). The three manufacturers that have commercially available models in North America are Ekolet, EcoTech, and Biorealis.

Following are example calculations of sizing (as promised) of two models — made by Ekolet and EcoTech. The concepts are applicable to all the carousel systems.

The Ekolet Koti (YV) Carousel Toilet

Cost: $$$$ ($5,000–$6,500)

The folks who make the Ekolet say it is designed for year-round use for 1–7 people. See Figure 6.7. (However, our own calculations suggest a maximum of 6.) It has a complete rotation cycle of three years (see below). A handy spray nozzle attached to the sink can be used to rinse and clean the commode as needed. Strengths are that there are no parts that can break or become blocked; it is well ventilated; it is functional during a

power outage; and the floor space required is minimal due to the tall vertical design. Additionally, due to the placement in the basement and the ventilation design, it offers another means of removing radon gas.

The required floor space is 1.3 m² (14 ft²), and 2.2 m (87 in) clearance is needed for height. The room needs to be heated, there must be a floor drain, there needs to be a waste drain for the leachate; and there needs to be enough room to provide access for the person tending to the composted materials. All the parts are small enough that they can be carried through a doorway. Power usage

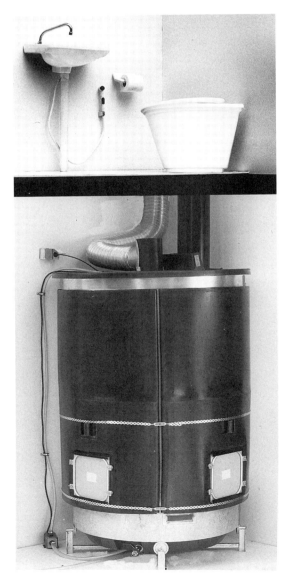

Fig. 6.7: *The Ekolet Koti YV compost toilet for up to seven full-time occupants is a carousel-style moldering toilet.*

Photo credit: Matti Ylösjoki, Ekolet Ltd, www.ekolet.com

is under 10 watts. Airflow is 15–30 m³/hour (9–17 cfm). Total volume is 2,000 L, divided between four 500 L compartments (500 L is the Container Size [CS] used in the equations below). One could opt to make a custom pedestal out of a suitable material (meeting the requirements noted in Chapter 3). The ventilation pipe size for the compartment is a 6" (160 mm) flexible ducting. Additional 4" (100 mm) venting can be connected at the back of the plastic pedestal if one wants to vent a bathroom that has a shower. This additional venting would be turned on just as a ceiling fan would, when having a shower, thus sucking the moisture from the air, down through the pedestal and outside.

Assuming that psychrophilic (very cold) temperatures do not exist, the three-year time frame for moldering would be expected to produce a sanitized material (as long as no new additions have been made) because it surpasses the time required at mesophilic temperatures for Ascaris ova (worm eggs) deactivation.

(*NOTE:* The Ekolet (YV) is comparable to EcoTech's large Carousel Composting Toilet.)

Though the manufacturer does not advise the use of bulking agents like sawdust, we recommend it. The unit itself is tall; without light and airy bulking agents, you increase the potential for compaction of the lower materials. It is highly recommended that you add 0.25 L (1 cup) per person per day.

CSs = Ekolet (YV)	500 L
E = Excreta volume	0.18 L/person/day
A = Additive volume (minimal, if any req'd)	0.5 L/person/day
O = Number of occupants	Unknown
H = Headspace needed	20 L
T = Time target for container replacement	365 days
Sf = Safety factor	Not required
Vr = Volume reduction %	50%
Vf = 1 − Vr	1 − 50% = 50%

Data needed to figure how many occupants (O) can be served by the Ekolet (YV)?

Calculation: *How many people can use this system (Occupancy [O] for Container Size [CS])?*

$CSs = (E + A)(O)(T)Vf + H$

$500 = (0.18 + 0.25)(O)(365)(50\%) + 20$

$500 = 0.43(O)365(50\%) + 20$

$500 = 78.48(O) + 20$

$500 - 20 = 78.48 (O)$

$O = 480 \div 78.48$

$O = 6.1$

Number of occupants that this system could service full time (figuring in the use of bulking agents) is a maximum of 6 regular users (one less than the maximum of 7 as recommended by the manufacturer).

(*NOTE:* If one incorporates red wiggler compost worms the finished volume after reduction would be less than 40% (if Vr% is > 60%), thus allowing the unit to service up to eight full-time users).

Ekolet Holiday Home (VS) Carousel Toilet

Cost: $$$$ ($4,000)

This Ekolet VS model (aka SISÄ) is similar in all respects to the Koti except the vent duct work is smaller, at 100 mm (4"), and the total volume capacity for its four compartments is 650 L. (A single compartment has a container size of 162 L; see Figure 6.8.) Based on a yearly rotation, this system could service a vacation home or be used full time in a home with two people. Once a compartment is filled, a skim of soil or peat is placed on top, and over the next three years, the compartment decomposes to decrease in volume 60–70% (to approximately 50 L). Unlike the larger system above, bulking agents can be avoided because the smaller volumes result in less compaction.

Assuming that psychrophilic (very cold) temperatures do not exist, the three-year time frame for moldering would be expected to produce a sanitized materials (as long as no new additions have been made) because it surpasses the time required at mesophilic temperatures for Ascaris ova (worm eggs) deactivation.

(*NOTE:* Ekolet's SISÄ model is comparable to EcoTech's medium Carousel Composting Toilet.)

Data needed to figure how many occupants (O) are served by the Ekolet VS: (See table right)

Calculation: *How many people can use this system (Occupancy [O] for Container Size [CS])?*

$CSs = (E + A)(O)(T)(Vf) + H$

$162 = (0.18 + 0.25)(O)(365)(40\%) + 8$

$162 = 0.43(O)365(40\%) + 8$

$162 = 62.78(O) + 8$

$162 - 8 = 62.78 (O)$

$O = 154 \div 62.78$

$O = 2.5$

This system, which was originally designed as a holiday/seasonal system, could potentially service two occupants full time if bulking agents were used. However, due to smaller bin height, compaction is less of a concern, so the bulking agent could be eliminated and worms could be used to increase the servicing capacity to up to four people, full time.

Drainage for any moldering system, including the carousel, is as simple as adding (18–25 cm (7–10 in) of untreated bark chips at the bottom, thereby avoiding the need

for screens and grates that come with a bin moldering system; the bark chips ultimately become part of the compost as they age. The use of bark chips is a growing trend in the waste treatment industry due to their effectiveness as a screen, but also in treating leachate and removing organic compounds that otherwise create high biochemical oxygen demand (BOD).

CSs = Ekolet VS	162 L
E = Excreta volume	0.18 L/person/day
A = Additive volume (if any req'd)	0.25 L/person/day
O = Number of occupants	Unknown
H = Headspace needed	8 L
T = Time target for container replacement	365 days
Sf = Safety factor	Not required, as highly serviceable
Vr = Volume reduction %	60%
Vf = 1 – Vr%	1 – 60% = 40%

Fig. 6.8: *Ekolet SISÄ VS is a carousel-type moldering toilet that is designed for vacation homes or two regular users.*
Photo credit:
Matti Ylösjoki, Ekolet Ltd,
www.ekolet.com

Moldering: Aquatron

Cost: $$$$–$$$$$ ($7,000–$10,000)

The Aquatron is an integrated carousel-type moldering toilet system developed in Sweden in the early 1990s — so they've been around a long time. The Swedish Environmental Protection Agency first certified the Aquatron in 1992, and today it has a CE International Certification and is internationally available. This system has also undergone the scrutiny of the International Living Building Institute's Declare process, which concluded that the Aquatron met the criteria to be used in Living Building Challenge projects. It doesn't get any greener than that…well, ok, maybe the Humanure bucket is greener.

The Aquatron system can take input either from full flush toilets or from a special urine diversion pedestal (called a *Dubbletten*). It removes the solids to be composted and diverts the fluids either directly to a sewer system or septic tank, or it moves the fluid through a UV sterilizer and a phosphorus trap (with a Polonite filter). The Polonite filter does not perform any sanitization; waters that travel through the UV sterilizer are sanitized. The sanitized water can be used for irrigation purposes, while the Polonite-filtered fluid is safe for use as sub-surface-discharged irrigation. The spent Polonite, once dried, is full of phosphorus, and can be used as a granular fertilizer applied by hand to surface soils. In short, it is very flexible in how it is used.

The separator design uses no power and has no mechanical moving parts; it just relies on gravity, the shape of the separator, and a special and simple screening system. It accepts flushed material, directing them into a vortex, where the solids drop out and fluid is separated out; it has a fluid removal efficiency of between 90%

to 98%. There are three sizes of separator; the choice is dependent on the number of toilets (the largest can handle 10 toilets).

The collection container that receives the solids comes in nine sizes, though the most likely size for residences will be the Aquatron 4×200 (which serves a home of five people full time) and 4×300 (which serves a home of ten people full time). Both of these house systems are carousel designs (see Figure 6.9). Aquatron also offers models with larger collection systems for public buildings (similar to the Clivus types discussed in Chapter 7), and a smaller sealed wheelie bin moldering system for cottages and cabins.

The manufacturer recommends that the carousel be rotated every 3–4 months, for a complete rotation of 12–16 months. Carbon bulking material is added through the top compartment on the bin as required to balance the C:N and soak up any excess moisture. Worms are commonly added to a bin, which greatly accelerates decomposition. From initial filling to the final aged and cured/matured stage, volume reduction is 90%.

If one wants to capture and use the fluids in greywater, then the liquids are passed through a UV sterilizer and then into a tank containing a media that traps phosphorus, removing 70% to 90% of it. The filter media also provides additional pathogen removal due to its high pH (Nilsson et al., 2013).

Polonite is a natural mineral found in Poland (the downside is the need to ship this mined product long distances). Polonite is calcium rich, meaning it has positively charged ions; this allows it to remove phosphorus, which has negatively charged ions. It does this at an astonishing rate: 50 kg of Polonite provides the same function as four metric tons of sand. The filter media self-sanitizes; when dried, it can be applied directly to fields as a slow-release fertilizer. It is expected that one 50 kg bag will be required every two years; it's time to change it out when the pH falls below 9. Alternatives to Polonite include a similar mined product from Ontario, Canada, called Sorbitive; it achieves similar removal rates (Balch et al., 2013).

Leachate

The collection bin requires a leachate drain to an approved leachate system, septic system, or sewerage system.

Urine diversion

Urine diversion from the seat could be directed to a greywater system, a storage collection system, or to a leachate, septic, or other sewerage system.

Source-separated water

- The separated water flows, if untreated, would be directed to a leachate, septic, or other sewerage system.

- Treated water flows, via UV sterilization and a phosphorus trap, could allow for direct ground infiltration dispersal, or they could be combined with greywater flows.
- Plumbing could be designed to allow diversion to a leachate, septic, or other sewerage system either as a form of seasonal diversion (away from greywater) or as an emergency diversion (failure of the UV light or lack of access to phosphorus trap filter media).

Toilet pedestal

- Accepts standard low flush toilets.
- Accepts the Dubbletten urine diversion 1 L flush.
- Requires a 3" (76 mm) sewage conveyance from toilet pedestal to the separator.

Insect control

There is none, as the various toilet pedestals (full flush or the urine-diverting 1 L flush) all have the standard water trap built in.

Fig. 6.9: *Aquatron system: Schematic showing collection assembly for a model suitable for a residential application.*
Illustration credit: Gord Baird

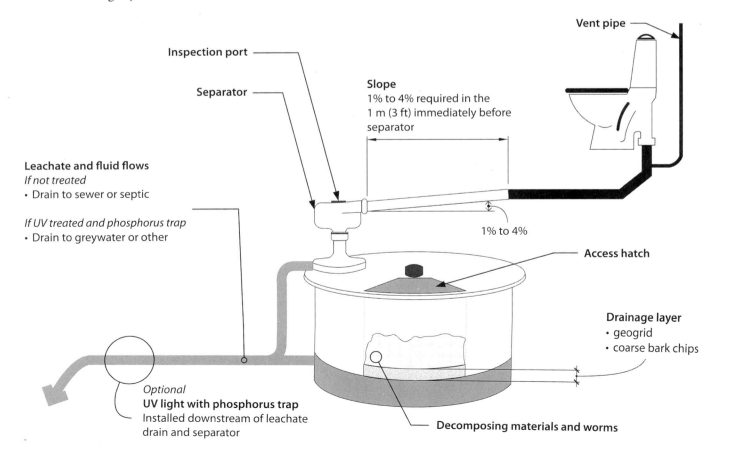

Vent pipe

Inspection port

Separator

Slope
1% to 4% required in the
1 m (3 ft) immediately before
separator

1% to 4%

Leachate and fluid flows
If not treated
- Drain to sewer or septic

If UV treated and phosphorus trap
- Drain to greywater or other

Access hatch

Drainage layer
- geogrid
- coarse bark chips

Optional
UV light with phosphorus trap
Installed downstream of leachate drain and separator

Decomposing materials and worms

Servicing

- Periodically view the top of the separator, accessed through the inspection cover on the top. Ensure there is no debris in the screen, and if there is, push it past the screen. If there is any scale or mineral build-up on the screen, the screen can be removed and soaked in CLR (Calcium Lime Rust remover) to clean it, or it can be replaced.
- On initial set-up of the bins, place 4–6 cm (1.5–2 in) of coarse bark chips (e.g. pine bark chips sold in garden stores) with a piece of geo-textile (thick filter cloth) over top, and then 2 cm (¾ in) of soil or compost.
- Inspect chambers once per month to ensure there is proper drainage; add carbon (wood chips, sawdust) if it is too moist.
- Add worms to aid in decomposition. This is optional.
- Rotate chamber as required (a quarter turn every 3–6 months).
- Change UV light when alarm goes on (every 3–5 years).
- Empty the longest-aged bin when required.
- If using a phosphorus trap, check the pH every 6 months; change the media when it's lower than pH 9.
- Clean fan and screen assembly yearly.

Power requirements

The UV system requires 35 watts; the fan needs 5 watts.

 Advantages:

- Allows the conversion of standard flush toilets to a resource recovery system. Often way less expensive than a whole new on-site septic system. Extends life of old septic system.
- Allows the flexibility to collect humus — and potentially phosphorus fertilizer — if so desired.
- The basic operation of solid/fluid separation and solid collection requires no power other than a fan.
- Fully treated fluids (if that option is chosen) can produce sanitized irrigation water.
- The UV-sterilizer lamps have a 3–5 year lifespan, and they come with an alarm that goes off when one burns out. The UV unit is rated at 35 watts.
- The emerging use of phosphorus trapping media is quickly gaining proponents, and more products are coming online for easier, more local access.

 Disadvantages:

- If water conservation is paramount, this system is unsuitable.
- The Polonite filter media is a consumable.
- The US distribution of the Aquatron systems was in transition at the time of writing, but it is suspected that this will resolve itself. Direct contact with the European head office will either arrange to put you in contact with the new North American distributor, or arrange shipping from Europe.
- Cost: $$$$–$$$$$ ($7,000–$10,000)

We are very interested in this system due to its ability to integrate with typical flush toilets and its long track record of use. Often the most ecological technologies derive from Europe; North America tends to lag about a decade behind. This technology has matured, and it has a place now as a viable option for retrofits and resource recovery.

Ecodomeo: Hybrid — Continuous Moldering

Cost: $$$$ ($4,500–$6,000)

The Ecodomeo compost toilet (Figure 6.10) combines features of the moldering systems just described with continuous systems (which will be discussed in detail in Chapter 7) — with one major difference. The Ecodomeo toilet simplifies the plumbing to handle various fluids. Most systems with urine diversion seats are tasked with managing both urine drainage *as well as* handling leachate that accumulates in the solids collection area. The Ecodomeo diverts all fluids away from the solids just inside the pedestal *without* using a urine-diversion seat. (Figure 6.11) This allows the solids to be conveyed to a storage system that does not require a leachate handling method. The Ecodomeo's special pedestal seat design is fail-safe, suiting the variety of behaviors that come with poor potty training — with any anatomy or age group.

The seat itself is small and can be retrofitted into existing homes with minimal work and cost. The conveyance of the solids can be adapted to several collection containment systems, from large containers to a dedicated exterior collection room. Additions to buildings can also be made to allow for easy collection, again an opportunity that allows for retrofit. See Figure 6.12.

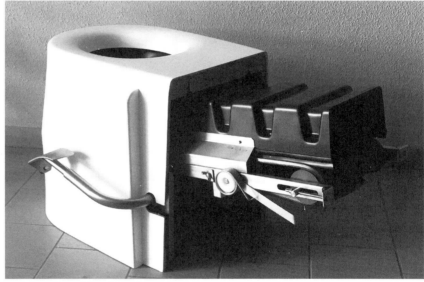

Fig. 6.10: *Ecodomeo*™ *pedestal assembly incorporates manual ratchet system to convey solids away.*

PHOTO CREDIT: EMMANUEL MORIN

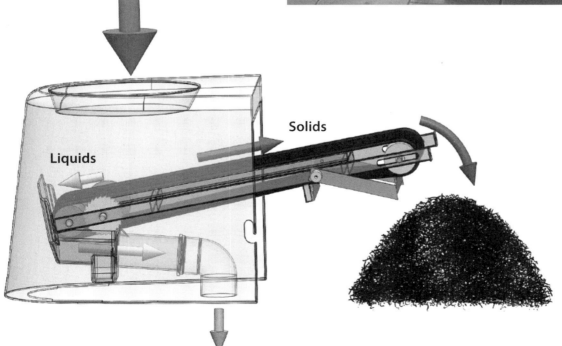

Liquids

Solids

Fig. 6.11: *Ecodomeo fluid and solids pathways.*

ILLUSTRATION CREDIT: ECODOMEO

Key Design Elements for the Ecodomeo Toilet

The pedestal is a stand-alone unit that separates urine and feces in a novel way. The pedestal has a conveyor belt that is tilted up toward the rear of the toilet; it extends out the back, into an adjacent collection chamber. Solids are transferred via this conveyor belt while allowing most fluids to drop to the front and exit down a drain. The conveyor is advanced by way of a foot pedal that employs a ratchet; the flow of materials are illustrated in Figure 6.11. The collection compartment has at least two mounds (sometimes a third); the active mound receives new additions, and the adjacent worm mound is decomposing and being digested. If a third mound is used, it is a sanitizing mound. As needed, the most mature mound is removed, and the subsequent mounds are shifted into the empty spot.

- Stand-alone pedestal.
- Requires a 2" drain connected to a leachate system.
- Collection chamber can be adjacent or below, or it can be bins (flexible application).

- No bulking agent required, but it can be used.
- Collection compartment is well sealed and vented to suck air in through the pedestal, across the mounds and out of the compartment.
- Two-story applications can be accommodated, as pictured in Figures 6.13 and 6.14.

Sizing

- Very flexible, as are all batch systems.
- Manufacturer suggests a floor area of 3 m² (32 ft²) enough room to have three piles.
- Collection area is best to be set a minimum of 40 cm (16 in) lower than the floor height that the pedestal sits on.

Servicing

Dedicated Compartment Collection — Annually, or even every 2–3 years, the compartment is accessed, and the cured materials are removed, to a sanitizing step if required or direct burial, depending on the time the materials have been held without new additions and on the state of maturity (nitrogen and state of decomposition). The active pile is moved

Fig. 6.12: *The Ecodomeo as installed on a single floor. Note that the collection area is dropped (preferably 40 cm [16 in]) from the height of the toilet pedestal to allow height for collected materials.*

ILLUSTRATION CREDIT: GORD BAIRD

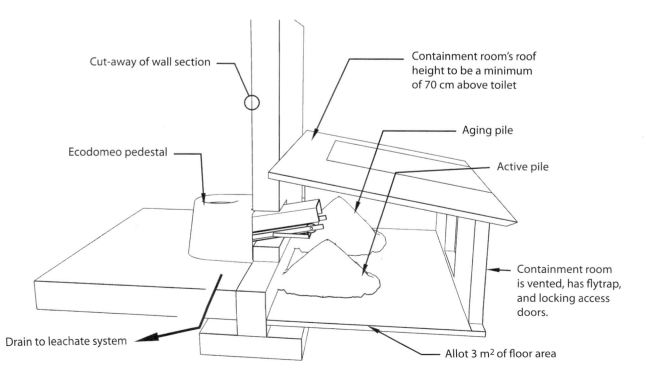

Cut-away of wall section

Ecodomeo pedestal

Drain to leachate system

Containment room's roof height to be a minimum of 70 cm above toilet

Aging pile

Active pile

Containment room is vented, has flytrap, and locking access doors.

Allot 3 m² of floor area

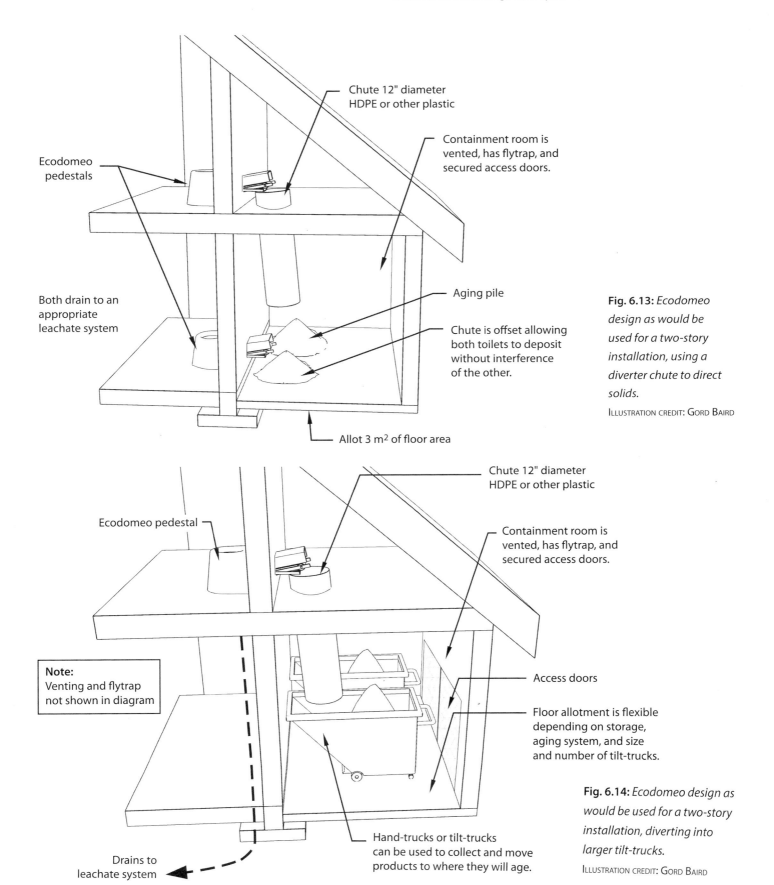

Chute 12" diameter HDPE or other plastic

Containment room is vented, has flytrap, and secured access doors.

Ecodomeo pedestals

Aging pile

Both drain to an appropriate leachate system

Chute is offset allowing both toilets to deposit without interference of the other.

Fig. 6.13: *Ecodomeo design as would be used for a two-story installation, using a diverter chute to direct solids.*

Illustration credit: Gord Baird

Allot 3 m² of floor area

Chute 12" diameter HDPE or other plastic

Ecodomeo pedestal

Containment room is vented, has flytrap, and secured access doors.

Note:
Venting and flytrap not shown in diagram

Access doors

Floor allotment is flexible depending on storage, aging system, and size and number of tilt-trucks.

Fig. 6.14: *Ecodomeo design as would be used for a two-story installation, diverting into larger tilt-trucks.*

Illustration credit: Gord Baird

Drains to leachate system

Hand-trucks or tilt-trucks can be used to collect and move products to where they will age.

over to the digesting curing area where worms can be reintroduced, allowing the now-empty collection area to receive new inputs.

Large mobile tilt-trucks as illustrated in Figure 6.14 provide for tidy and contained collection as an alternative. Coming in sizes from 0.5 m³ (132 gal) to 1.5 m³ (396 gal), compartment rooms are sized to hold two to three tilt-trucks.

Multiple Bathrooms

• Retrofits require special attention to carpentry and cabinetry.

• Chutes from upper bathrooms need to be tilted to deposit materials just behind lower unit (Figure 6.13, Figure 6.14).

Cleaning

• Regular cleaning of toilet seat, as for any toilet system.
• This system can be cleaned and rinsed with waters which flow through to the leachate system.
• A water faucet with flexible spray nozzle is recommended to be installed close to the toilet for easy cleaning; remote locations could have buckets of water poured in to perform the same function.

Other

• Insect control can be managed with use of screening over the exhaust fan.
• Fly control is aided by use of a flytrap that comes with the Ecodomeo unit.
• Optional fly control systems can be made as illustrated in Figure 6.15 and Figure 6.16.

👍 *Advantages:*

• The system offers some flexibility to install into pre-existing homes with proper

Fig. 6.15: *Flytrap for interior location inside the collection area: A pipe penetrates the wall, flies enter following the light from outside and the odor that migrates through the fabric bag; they get trapped and die.*
ILLUSTRATION CREDIT: GORD BAIRD

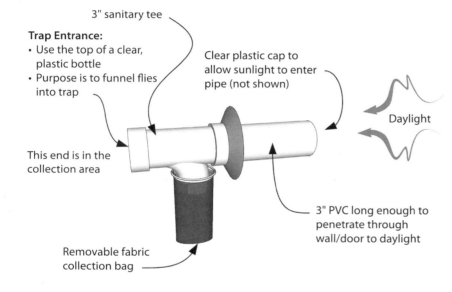

3" sanitary tee

Trap Entrance:
• Use the top of a clear, plastic bottle
• Purpose is to funnel flies into trap

This end is in the collection area

Clear plastic cap to allow sunlight to enter pipe (not shown)

Daylight

3" PVC long enough to penetrate through wall/door to daylight

Removable fabric collection bag

Fig. 6.16: *Flytrap: Entranceway is made of a clear plastic bottle top, cut off and inserted into the pipe.*
ILLUSTRATION CREDIT: GORD BAIRD

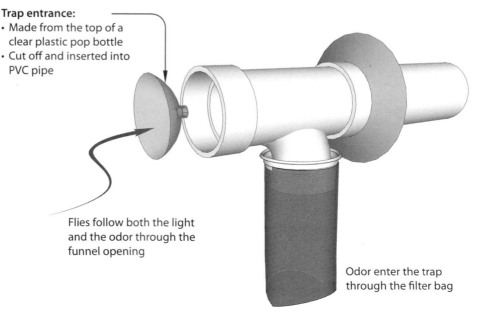

Trap entrance:
• Made from the top of a clear plastic pop bottle
• Cut off and inserted into PVC pipe

Flies follow both the light and the odor through the funnel opening

Odor enter the trap through the filter bag

carpentry to ensure a collection room is properly sealed and vented, and easily accessible.

- Because there is no bulking agent required, the collection chamber can be smaller than that of other continuous chamber systems.
- The lack of fluids results in less odor production. The use of worms in early-stage stabilization speeds up pathogen death, though helminth ova are not affected by the worms' digestive processes, so time frames for sanitizing are the same as noted in Chapter 2.
- These systems are well suited to remote locations because of the limited infrastructure required to be shipped, the ability to build an appropriate storage collection chamber on-site, and the lack of having to import bulking agent.
- Ideal in high-volume or public installations.
- Use minimal electricity (only a fan).
- Use minimal water.
- Minimal maintenance.
- The systems have another strength in that they process materials quicker, and there is less ammonia to impact the ecology when finished materials are reintroduced.
- Cost: $$$$ ($4,500–$6,000), depending on seat choice and conveyor belt length.

 Disadvantages:

- The conveyor belt does require cleaning, so despite materials being conveyed away from sight, residues are left, and undoubtedly bouts of diarrhea will be unsightly and require immediate attention for aesthetics.
- Some of the remote installations visited by the author relied on passive ventilation and were prone to minor odor. Active ventilation is highly recommended, as with any toilet.

Moldering: Other Systems
BioLet 30NE

Cost: $$$ ($2,000)

BioLet 30NE is a relatively simple, self-contained manufactured system that has a collection bin that sits inside the pedestal. When full, the same bin is set outside to continue to molder (age). It comes with a 12V DC fan and is a useful option for an instant set-up/install in a cottage or vacation home. With additional bins, as with any batch-type compost toilet, the system can easily expand; with more bins, the system can handle a surge in peak usage and therefore could be used in a full-time residence.

EcoTech Carousel

Cost: $$$–$$$$ ($3,000–$5,000)

Referred to already, this carousel system is the US-manufactured version of the original carousel designed by Vira Miljø, and, in all respects, it matches the Ekolet carousel system already discussed in detail. EcoTech is owned and operated by David del Porto and Carol Steinfeld, two very influential people in the industry of compost toilets and resource recovery. They have been involved in hands-on design, installs, policy and education, and are the authors of the *Composting Toilet Systems Book* published in 1999.

The EcoTech Carousel is offered in two sizes, medium (servicing 4 people full time) and large (servicing 6 people full time). For larger volume usage, accommodations can be made by contacting the manufacturer directly.

Biorealis Systems Inc.

Cost: $$$ ($500–$1,500)

Biorealis Systems, located in Alaska, builds compost toilet systems and sells the plans for a homemade carousel-style moldering compost system. This system looks to meet all the quality and criteria for a safe, functioning, and practical alternative. DIY plans are available at http://biorealis.com/composter/rotating.

Incineration Toilets

Cost: $$$$$ ($5,000–$7,000)

A system unto itself, incineration is not a form of composting, but it can be used in conjunction with moldering depending on seasonality. In very cold climates where access to a composting processor is limited during the winter months, incineration toilets offer the option of collecting and incinerating the solids. Some of these toilet systems, like those manufactured by Separett, allow the collected materials to be either composted or incinerated (so, they are flexible systems). In a cold climate, where an external compost bin may not be accessible for 1–4 months, incineration could be relied upon until temperatures change and disposal into the compost pile can restart. Other incineration schemes are all-in-one.

Separett makes two products that employ incineration. The first, a small incinerator called the Burn 900, is fueled by wood (1.5 kg per burn [3.3 lbs]), that is 36 cm (14 in) wide, 60 cm (24 in) tall, and it has a tiny chimney that is 2.2 m (7.2 ft) high. With this system, you would collect your solids in one of their bagged collection toilets (e.g. Villa 9220), then deposit the bag into the incinerator as required. This little incinerator is kept outside, under cover; it is fired up as needed. Materials remaining are ash, which is sterile. A household of two may need to burn once every three weeks.

The second Separett product is the Flame 8000, which is a toilet that offers incineration within the toilet itself.

ECOJOHN makes two complete urine-diverting toilet units that incinerate the solids, offering no opportunity for composting; both models rely on fossil fuels for the incineration. Either 12V DC or 120V AC is used to run an auger to convey materials to the burn chamber, and then it is ignited. The ash is sterile. Used in remote camps. Expensive.

Incinolet has an all-in-one unit that does not require urine diversion. It relies strictly on electricity (1.5 kW per cycle). Like the ECOJOHN models, it has its application — if no other options for composting exist.

Chapter 7

Continuous Systems

THERE ARE TWO SUBCATEGORIES of continuous systems: all-in-one units, and large centralized two-piece units consisting of the toilet pedestal and a separate chamber. They are very distinct from one another, but both rely on similar concepts.

Self-Contained All-in-One

These are probably the most common systems people envision when they think of compost toilets: a large box with a toilet seat, perhaps a step-stool, drawers, knobs, and handles. The benefit of a unit that houses everything — the toilet pedestal, receptacles, fans, drainage, and mechanical components — is that they can be installed in any room, with limited alterations. (Renovations are still required for installing vent pipes and leachate drains, however.) Due to their size, all-in-one systems tend to be designed for seasonal/recreational use only, or for a very limited number of regular users.

We recommend that readers exercise their due diligence before choosing a self-contained system. These systems do have a place, but we urge caution (except with systems like Nature's Head or Air Head). As part of the writing of this book, we prepared a questionnaire for people living with compost toilets, asking them to share their experiences. Negative responses were primarily around the issues with self-contained units; people had complaints about clogging leachate drains, the caking of solids that had to be physically cleaned, flies, and the requirement to remove raw feces to perform these cleaning tasks. On the other hand, some people using these systems had no problems and loved their all-in-one system.

One of the inconveniences of self-contained systems is their limited capacity to handle sudden surges in usage. In order to correctly size a self-contained system, it is imperative to think about peak usage — and perhaps to choose a model one size larger than seems needed.

Design considerations include planning for either AC or DC for a fan and optional heating element, planning ventilation piping to the exterior of the building, and ensuring that any urine diversion or leachate drainage has the proper drop (slope) to function.

For all continuous systems, a consideration is that temperatures should be maintained between 18°C–31°C (64°F–88°F) for the microbes to stay active; below 13°C (55°F), the compost process will slow or stop altogether resulting in a much faster fill rate.

Before buying any system, ask about warranties and accessibility to spare parts. View the spare parts list and read the service manual so you know what to expect in terms of cleaning and maintenance.

There are many products out there, and bound to be more. Following is list of some of the most popular self-contained units, several of which we have seen in action, one we have owned (but replaced with our present system), and a couple that our clients have had experience with.

BioLet Models

Cost: $$$–$$$$ ($2,500–$4,000)

BioLet is a company that makes multiple versions of self-contained continuous systems ranging from a four-full-time user model (BioLet 65) to much smaller capacity toilets (BioLet 25, BioLet 15). (*NOTE:* BioLet is the international brand name. The same models are called EcoLet in Australia, and MullToa in Sweden, where they were originally developed.)

These models have a maximum power draw of 355 watts, though on average they use just 55 watts. Caution must be taken in cold-climate applications, where homes may be left unheated and the contents can freeze; if the unit is turned on before things thaw out, the motor will turn and shear off the shear pins.

These models are not completely trouble-free. There are reports of materials not staying equally moist, and thus there is a need to add additional moisture into the rear of the unit, especially if one is away for more than two days; excessive drying will lead to the shear pins breaking. The BioLet 65 has received the SWAN Ecolabel, a prestigious Scandinavian certification.

On a regular service schedule, expected emptying is every three weeks; by then, the materials are only marginally stabilized, so care must be taken when moving materials to a composting area for curing and sanitization before dispersal.

Sun-Mar Models

Cost: $$$ ($1,500–$3,000)

Sun-Mar has a long history with the manufacture of self-contained units, and their models are perhaps the most widely sold brand in North America. The company has extensive documentation on their toilets, which are vast and varied. They offer both on- and off-grid units that rely on the rotating of a drum to separate solids from liquids and mix new with old; in the process, solids are deposited into a tray. All their units can be equipped with heating elements.

Their centralized models (Centrex) seem to be more consistent. Centralized systems will be discussed in the next section of this chapter, but to introduce the concept, they are a stand-alone central processing that has one or more pedestals fed into it.

Reading customer reviews will help you, as they are highly variable and offer insight into how to manage this type of toilet. In our experience, they are best suited for low-volume consistent use wherein materials do not have a chance to dry out.

Nature's Head

Cost: $$ ($1,000–$1,500)

Nature's Head composting toilet is flexible in that it can be a batch system, a moldering system, and/or a continuous system. It employs urine separation that is easily accessible and easy to maintain, allowing solids to be collected in the dedicated chamber attached to the seat unit. It consistently ranks high in customer reviews. Originally developed by a couple of sailors, the Nature's Head quickly become recognized as a sturdy, well-built, simple unit useful in homes and tiny houses. Urine is collected in a front storage tank; the tank requires frequent emptying, but that task is fairly simple. Based on a 2.2 gallon storage tank (8.3 L), two regular users would fill up this tank every 2–3 days.

The Nature's Head is rated for 60–80 uses before the solids compartment needs to be emptied; for two people, that equates to every 2–3 weeks. The process is a little cumbersome as it involves unclipping the top half of the toilet to gain access to the collection chamber, and then dumping that chamber into a compost pile for further processing. You can also just replace the collection chamber with an empty one. You can purchase these as additional items. They come with lids, so storing and batch moldering can meet the needs of higher-use homes.

If the unit is left too long without new additions and regular handle turning (there is no motor), materials can dry and stiffen. Adding water, coffee grounds, or moist peat will help break up and loosen the materials. In the event the materials have completely hardened, then the contents of the compartment need to be removed or the bin swapped if you have a

spare bin while leaving the full one to molder, or soften, then be dumped and composted. All aspects of the design allow flexible handling and not much chance of failure.

Removing the *struvite,* a mineral that builds up in the urine bottle, is as simple as dumping the bottle in a compost pile or under a tree, adding some pea gravel and vinegar, shaking, and emptying.

See Figures 7.1 and 7.2.

Air Head

Cost: $$ ($1,000–$2,000)

Air Head is a competing brand of the Nature's Head toilet; it is designed around the same principles.

Envirolet

Cost: $$–$$$ ($1,500–$3,000)

Envirolet offers several self-contained models that are rated for 4–6 full-time users; they are offered in both 12V DC and 120V AC (there is also a centralized version like

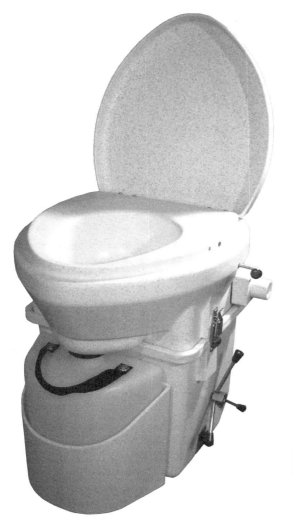

Fig. 7.1 (left): *Nature's Head Toilet.* PHOTO CREDIT: M. MILLER AND L. STEARNS, NATURE'S HEAD

Fig. 7.2 (below): *Nature's Head toilet access.* PHOTO CREDIT: M. MILLER AND L. STEARNS, NATURE'S HEAD

the Sun-Mar). These models have their own mixing mechanism inside, which employs a screened grate and agitators. These toilets can be used with a heating element, which can be switched on/off separately from the fan. The power requirement is 540 watts, with average use of 40 watts to run the fan.

The Envirolet suffers from some of the same problems as the BioLet: a conditioned space where contents do not freeze is important, and there is a tendency for materials in sections not immediately under the seat to become overly dry and require an extra hit of moisture. Ideally, like the Sun-Mar, continued light use is best. When materials are left undisturbed for extended times, they can become hardened and block the grate. More info can be found at www.envirolet.ca and www.envirolet.com.

Key Considerations for Self-Contained Systems

- Before purchasing a system, read customer reviews; review the user manual of any system you are considering, paying close attention to the troubleshooting section.
- Realize that materials will require composting outside of the toilet system for them to be considered cured, let alone sanitized.

- Plan for leachate handling systems; or, in the case of the Nature's Head, a method to use, dispose of, or store urine until it is required.
- Slab-on-grade installations do not lend themselves to easy drainage of leachate; in those situations, you should install where there is access to appropriate built-in drainage potential (slope) thus avoiding the need for using inappropriate shallow containers (e.g. cake pans or pie plates) to collect fluids in.
- Easy to retrofit.

Ideal for:

- Cottages, cabins, remote areas.
- Some are ideal for boats, tiny homes (Nature's Head, Air Head).

Not Ideal for:

- Buildings that are left unheated and will freeze in the winter.
- On-grade bathrooms that do not allow for an effective drainage hook-up (an exception is the Nature's Head toilet).

Centralized Continuous: The Big Systems

Continuous systems are those that allow a high degree of decomposition and processing, allowing for "finished" materials removal while continuing to accept new raw inputs. Centralized continuous manage in part due to their size and the transit time for materials to migrate from the entry to the exit; they are a varied group of products, usually manufactured. They serve the need of processing larger volumes, allow connection of multiple toilets, allow the use of various conveyance systems (like foam-flush or vacuum flush), and require less frequent servicing. Larger volume processing means large storage chambers; large chambers require some form of mechanical processor that turns, scrapes, or agitates to screen materials to promote aeration and the settling of more decomposed materials to a final collection point.

Flush versus No-Flush

When flushing is used as a transport system, multiple toilets in different locations and on different floors can be serviced by a centralized compost toilet reactor/chamber. Gravity is required for these to work unless a vacuum system is used.

No-flush designs, often called *direct drop*, must sit directly above the chamber and be connected by chutes. These are limited to a maximum of two toilet pedestals per commercially available processor, though it's most common to have a single pedestal. Some toilets use a long direct-drop chute from a second floor to a basement system, but we do not recommend this arrangement; the sides of the chutes inevitably collect solids (the brown kind), which increases the opportunity for the yuck factor and unpleasant maintenance. The shorter the drop, the less scrubbing of stuck-on plop.

The processing chambers all have their own way of handling and processing materials; we will expand on this below, but, in essence, they continue to receive new raw materials while older materials decompose and migrate to an access port where they are then usually considered stabilized and ready for removal to a compost pile. Due to the continuous addition of raw materials, even highly stabilized materials are to be considered uncured and highly pathogenic — despite manufacturer's claims.

You might ask why a material two or three years old is not matured/cured? Going back to the discussion on composting in Chapter 2, maturation is characterized by a reduction of phytotoxic compounds, (e.g. volatile nitrogen and ammonia), and other compounds toxic to humans; matured material will not impact available soil nitrogen when introduced into the soil. With new nitrogen-rich fluids entering and draining into the completed materials, those "finished" materials become replenished with the items we want to have removed. For this reason, collected materials should always be further composted to reduce pathogen loads and phytotoxic compounds before dispersal. Centralized systems are ultimately the most complex and expensive options, but given the variety of conveyance options (different flush options), they can fit into retrofits nicely.

Key Considerations for Centralized Continuous Systems

- Cost: $$$$$ ($7,000–$15,000+)
- Space: Most continuous systems are located in a basement or a crawlspace tall enough to accommodate the storage unit; this allows various gravity-fed conveyances

to them. However, if a vacuum conveyance method is used, chamber placement is irrelevant. The space housing the storage unit needs to have enough room to allow you to work with rakes or other long-handled tools.

- Power: All these systems require fans; some have motors, pumps, and heating elements.
- Heated space is a must for units installed in climatic zones 7B and colder.
- Professionals are likely to be involved in wiring, plumbing, and carpentry.

Fig. 7.3: *Inclined plane concept drawing associated with Clivus designs.*

ILLUSTRATION CREDIT: GORD BAIRD

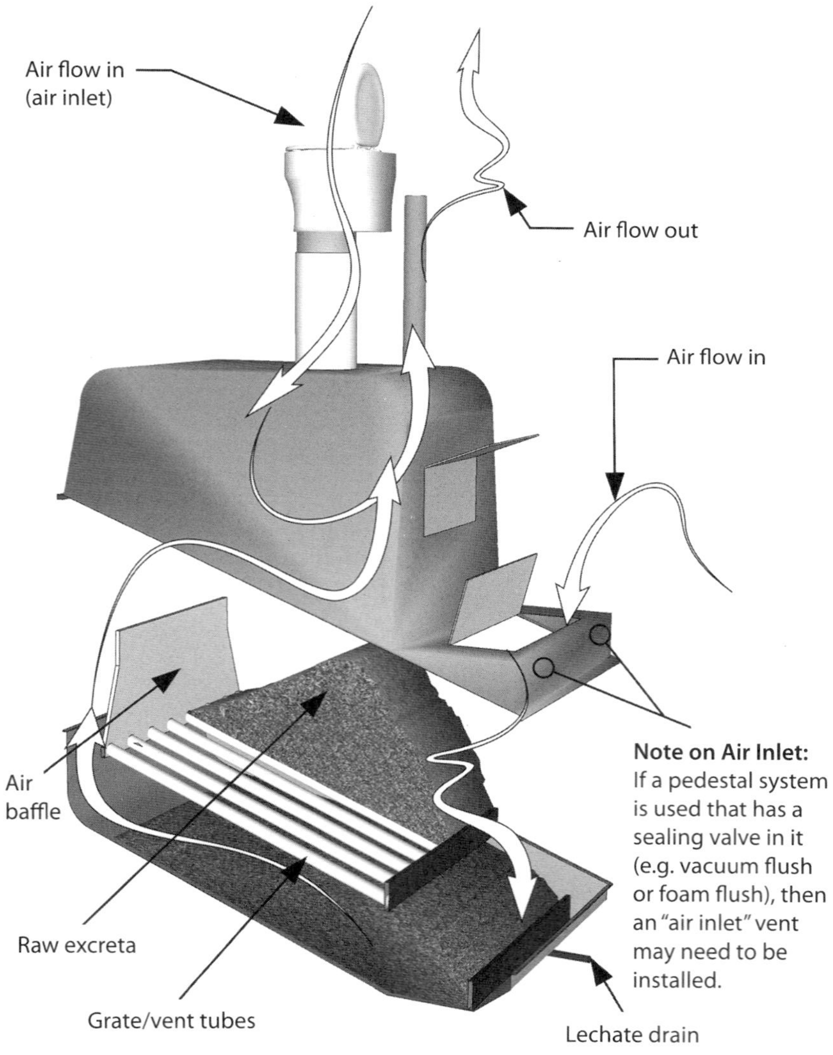

Air flow in (air inlet)

Air flow out

Air flow in

Air baffle

Raw excreta

Grate/vent tubes

Note on Air Inlet: If a pedestal system is used that has a sealing valve in it (e.g. vacuum flush or foam flush), then an "air inlet" vent may need to be installed.

Lechate drain

- Floors need to be designed to hold the weight of the chamber; therefore, if not on a slab, a structural engineer may be needed to confirm or design a floor system that is appropriate.

Ideal for:

- Homeowners that want a flush toilet for aesthetics or more flexible bathroom layout
- Homes that have larger capacity needs
- Homes with more than one toilet
- Retrofits
- Single-floor buildings where toilets and collector are on the same floor (if using a jet vacuum)

Not Ideal for:

- Cabins, cottages, or small outbuildings — due to size, costs, and the availability of simpler systems that meet the same needs

Clivus (Inclined Plane Styles)

Cost: $$$$$ ($7,000–$20,000)

The Clivus Multrum, is the granddaddy of continuous design (see Figures 7.3, 7.4, and 7.5). Any time a product is used for a Living Building Challenge (LBC) project, it deserves special mention: the Chesapeake Bay Foundation's Brock Environmental Center uses the Clivus Multrum and is a fully certified LBC project.

Clivus-type systems are often referred to as *inclined plane* systems. As they receive new materials, these are deposited on a suspended grate system, and accumulate over time. The lowest (or oldest) deposits decompose and fall through the grate to an inclined plane and migrate down to the lowest access hatch where (after several years), they can be collected and used in the landscape. Air flow and hatch access are illustrated in Figure 7.3

and Figure 7.4. These systems all require a leachate drainage system.

Periodically, the rear upper inspection hatch must be opened so that mounded materials below the pedestal can be raked and flattened out. This action also helps break up hardening deposits that otherwise could completely crust over and block the grate system.

There have been many versions of this Clivus "inclined plane" developed by other manufacturers and DIY builders, with

variations on the main components of slope, grates, and drainage.

Though concrete has been used for constructing site-built chambers, it has turned out to be problematic: the surface of the concrete becomes pitted from the humic acids, urea, uric acid, and precipitation of urine deposits, thus impairing the migration of materials down the incline. This issue can be reduced by the application of acid-resistant sealers such as acid-resistant mortars, the silicone-based (NSF 151 rated) nontoxic

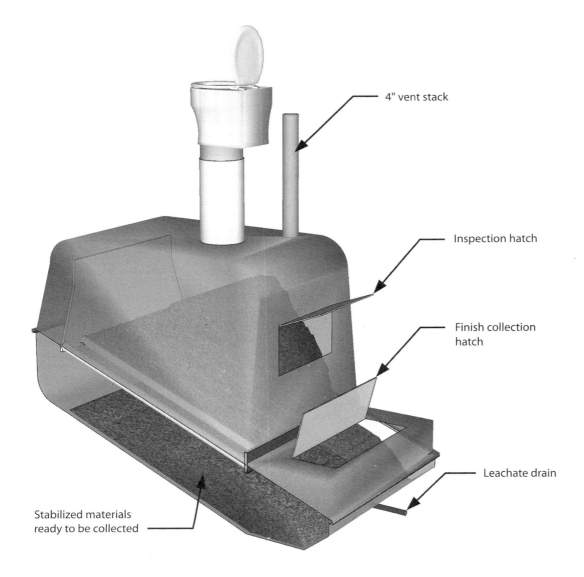

4" vent stack

Inspection hatch

Finish collection hatch

Leachate drain

Stabilized materials ready to be collected

Fig. 7.4: *Hatch access ports for Clivus-like systems.*

ILLUSTRATION CREDIT: GORD BAIRD

Fig. 7.5: *Clivus Multrum compost processor installation by Chris Magwood and the Endeavour Centre.*
Photo credit: Chris Magwood

¼" thick 3" × 3" angle iron, or schedule 80 steel pipe or schedule 80 PVC 3 pipe spaced 20–30 cm (8–12 in) apart. For the initial set-up, the grate system needs a temporary material laid on top of it so the first batches of excreta will not fall through to the inclined plane before processing can occur. This temporary material could be a piece of cardboard topped with a thin layer of compost, or, in place of the cardboard, wafers of straw or hay could be used.

The Clivus Multrum systems in Europe continue to be modified and improved. A new line of processors have been introduced that do not rely on a grate to separate finished materials from those undergoing decomposition. Instead, they have a drainage barrier that allows fluids to pass through to a lower collection tank; the same barrier is also the inclined plane that directs finished materials to the exit. This design change addresses one of the weaknesses in the older inclined-plane models where excess fluids could build up and saturate the "finished" materials. The updated European models are not available in North America yet — possibly due to present contractual and distribution arrangements. Eventually, we expect to see these improved and updated Clivus models in North America.

The use of bulking agents (sawdust or wood chips) is strongly recommended to provide a coarse structure that will help avoid compaction. The manufacturer recommends adding 2 L (half gallon) of bulking agent per ten uses, which comes close to being the same amount used in the commode batch systems.

Inclined-plane-style systems have been incorporated successfully into public buildings, such as the Vermont Law School and the Bronx Zoo. However, they have been

sealers used on roofs for rainwater collection, or, in some cases, epoxy sealers.

Slope of the inclined plane varies between 20° and 30°. The grate is also sloped and terminates at the back against a baffle, which blocks new raw undecomposed materials from entering the finished material removal area.

The grate has to be strong enough to support several hundred pounds of excrement. This is achieved by using products like

used *unsuccessfully* at the University of British Columbia's C.K. Choi Building and several other highly publicized facilities. Due to the varying degrees of success, it is worth connecting with folks at facilities that use and promote them to hear/see how they service and manage their systems.

Sizing

If you are considering purchasing a manufactured system, the manufacturer/retailer will offer you size and design options depending on usage patterns — ranging from seasonal use to daily residential use to the heavy usage of a commercial or institutional building. Using the sizing equations given here, it is wise to perform your own calculations to verify a manufacturer's claims and see if your needs can really be met. The table in Figure 4.3 in Chapter 4 may be a handy reference as you work through the example.

Below, we will assess what size of processor we would need for 5 people, and then look at what Clivus Multrum recommends. We can compare our size to that of the recommended model and make sure they match.

Data needed to find correct Container Size to accommodate five full-time users (CS) (See table right).

(*NOTE:* Volume reduction is approximately 40% over one year; therefore, Volume final (Vf%) is represented as 1 − Vr%, for a Volume final % of 60%.)

Calculation: *What should Container Size (CS) be?*

$CS = ((E + A) (O) (T) (Vf)) (1+Sf)$

$CS = ((0.18 + 1.0)(5)(365) (60\%)) (1 +40\%)$

$CS = ((1.18) (5) (365) (0.60)) (1.40)$

$CS = (1292) (1.40)$

$CS = 1809$ L

Using the above equation, we can determine the desired volume of a processing unit. When compared to Clivus Multrum's North American system for 4–5 regular users, the CM8 Model, we see it has a total volume of 2 m^3 (or 8,000 visits per year, 0.45 m^3 per person per year). Their recommendation matches our assessment.

(*NOTE:* The European version, the M150 servicing 3–4 users per year [or 7,000 visits per year, 0.29 m^3 per person per year], demonstrates that it has a higher rate of volume reduction or decomposition than the older North American version.)

Additional conveyance methods

Because fluids are able to leach through the pile and drain out through a leachate system, additional moisture is not too much of a concern in these systems. For this reason, the same processor can accept inputs from micro-flush, vacuum-flush, and foam-flush pedestals. These different pedestals types can be purchased separately from the processor and from the various distributors. Clivus Multrum's micro-flush and vacuum-flush use approximately 0.5 L (2 cups) of water. The use of these different flush systems allows for greater flexibility in toilet pedestal placement. Because of this flexibility, it opens the opportunity for these systems to work in retrofits, but only when the conveyance pipes can be ensured to meet the minimum

E = Excreta volume	0.18 L/person/day
A = Additive volume (strongly advised)	1.0 L/person/day
O = Number of occupants	5 people
Sf = 40% (headspace)	40%
T = Time target for container replacement	365 days
Vr = Volume reduction over time	40%
Vf = 1 − Vr%	1 − 40% = 60%

slope required, which is much greater than in standard building/plumbing codes.

Urine collection

Urine collection can be accommodated, but it is not common, as it is desirable to have moisture added to maintain some moisture in the top of the pile.

Leachate handling

Leachate handling is required with a Clivus inclined plane system. Leachate is collected through a 1½" drain (preferably) and sent to a properly designed leachate handling system (discussed in more detail in Chapter 8).

Cleaning

Beyond the routine cleaning a regular toilet requires, cleaning could require scrubbing direct-drop chutes or rinsing pedestal that have narrowed drains (jet-vac), because rinse water can't do a thorough job in such drains.

As in all compost toilet systems, avoid using ammonia, chlorine, or other harsh anti-life products; it is best to stick with the vinegars or hydrogen peroxide solutions.

Servicing

A big selling feature of the inclined plane systems is that they require very little servicing. The main task is opening the upper "inspection hatch" on a regular basis (monthly) and using a dedicated muck rake to flatten and adjust the pile in order ensure the airspace between the pile and base of the chute is no less than 30 cm (1 foot).

On a yearly schedule, the lower access hatch will be opened and finished materials removed. Realize that these "finished" materials are stabilized only, not sanitized. Before you apply your "finished" materials, they will have to be put through a process to cure and sanitize them.

Additionally:

- Use 2 L of bulking agent per 10 uses (250 ml or 1 cup after each use) either through the pedestal for direct drop systems, or through the back inspection hatch for other conveyance systems. Large systems compact; therefore, this material will help avoid the creation of anaerobic conditions. It will also help balance the low C:N ratio in the feces.
- Add compost-enhancing bacteria if heavy usage is expected or is occurring.
- Check pumps and drains to ensure all are working.
- Clean vent screens and fans yearly.

Insect control

Like all other systems, screened intakes and outlets (that can be removed and cleaned) and fans strong enough to vent odors are critical. If leachate drainage slows due to mineral or suspended solids in the drain port, opportunities will increase for insects. Keep the drains clear and use 1½" diameter pipe, if possible.

Making and installing a flytrap is advisable. Insects that hatch will be drawn to light sources. If the pedestal does not have a seal or privacy flap in the bowl, then there is potential for insects to be drawn from the opening and into your home. Flies do not make good house pets.

A flytrap, as pictured in Figure 7.6, incorporates a toilet flange that is attached to the side wall of the plastic tank (with bolts and washers) with a bead of sealer squished between. Inserted into the flange is a funnel (which can be a top of a plastic bottle cut off and glued in, as pictured). Attached to the flange is a sanitary Tee that has a clear lens on the end, and a glass jar below. The glass jar allows visual inspection; if so desired, it can have an attractant inside. The glass jar also allows light to enter and attract insects. The glass jar can be attached using a rubber coupler.

Power

Fans associated with these systems are usually 5 watts. Expect to run a substantial 20 cfm fan to keep the system functioning well.

These systems are best situated indoors in conditioned (heated) spaces. If they are located in unheated spaces in cooler climates (≤ Zone 7), electrical power for heat may need to be accounted for. In warmer climates (≥ Zone 8), consideration of electricity for supplemental heat is not a concern (conditioned space or not).

 Advantages:

- Decreased servicing.
- Enclosed initial containment aids in controlling insects.

- Can meet the needs of a variety of residential settings, from retrofits to small lots.
- Holds a large volume.

 Disadvantages:

- Prone to compaction due to large volumes.
- Older models can be prone to excess moisture in the "finishing chamber" (newer European models resolve this).
- The grate systems are prone to clogging and require raking raw materials, therefore despite the long periods between removing stabilized materials, there is still involvement with raw feces — likely, a monthly task.
- Cost: $$$$–$$$$$ ($7,000–$20,000). Costs are variable due to variety of options for flush and conveyance systems.
- Only recommended in warm climates (Zone 8 and higher).

Fig. 7.6: *External access flytrap for closed contained continuous compost systems (Phoenix and Clivus).*
Illustration credit: Gord Baird

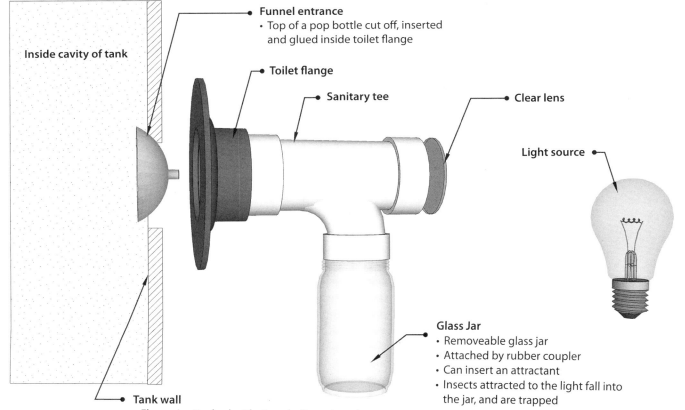

Inside cavity of tank

Funnel entrance
· Top of a pop bottle cut off, inserted and glued inside toilet flange

Toilet flange

Sanitary tee

Clear lens

Light source

Glass Jar
· Removeable glass jar
· Attached by rubber coupler
· Can insert an attractant
· Insects attracted to the light fall into the jar, and are trapped

Tank wall
· Flange is attached with stove bolts and washers
· A bead of sealer is placed between both before bolting together

Phoenix

Cost: $$$$$ ($7,000–$20,000)

The Phoenix has had a very successful history of use. Probably one of its biggest endorsements is having been incorporated into two Living Building Challenge projects: one in a residence (Heron Hall in Seattle), and one in a commercial building (Bullitt Center in Seattle). The Phoenix can accept a variety of conveyance systems, but they are predominantly installed with direct drop for residences, foam-flush for gravity conveyance, and the Jets vacuum pump. Unlike the Clivus Multrum, which relies on visual inspection and physical redistribution of raw materials as they build up, the Phoenix has tines inside the processor that can be turned from the exterior, removing the need to open up the processor unit on a regular basis.

Fig. 7.7: *Phoenix processor with a direct-deposit pedestal.* Photo credit: Michael Kerfoot, Sunergy Systems Ltd., co-designer of the Phoenix

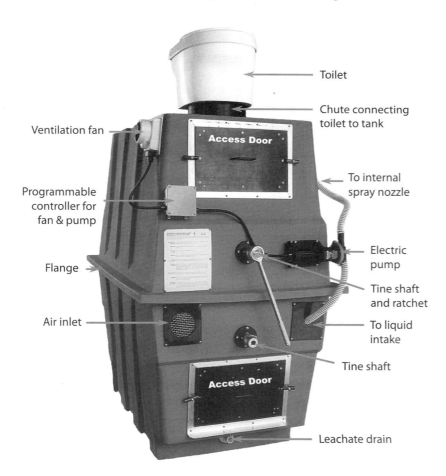

- Toilet
- Chute connecting toilet to tank
- Ventilation fan
- Access Door
- To internal spray nozzle
- Programmable controller for fan & pump
- Electric pump
- Flange
- Tine shaft and ratchet
- Air inlet
- To liquid intake
- Tine shaft
- Access Door
- Leachate drain

These processors can be sized to meet varying demand capacities. Higher capacity is accommodated by larger (taller) storage volumes and additional banks of tines. See Figures 7.7 and 7.8.

A single unit has a volume of 2.75 m^3 (3.6 yd^3) and handles 4–5 regular users. It maintains about ⅓ of the total volume for air space between the top and the first set of tines. Air baffles are built into the side wall of the processor to offer greater air exchange, and spray mist nozzles allow for the misting of moisture to the top of the materials to aid in maintaining beneficial moisture levels across the materials to avoid caking and promoting microbial decomposition.

Bulking agents are required; the *Phoenix Composting Toilet System Instructions for Operation and Maintenance* manual has this to say about bulking agents:

> The best bulking material is dry planer shavings from a white softwood such as pine. Do *not* use shavings from decay-resistant woods such as cedar or redwood: this material will *reduce* the composting rate. The bulking agent must have a physical structure that resists compaction so that air voids will remain open. Do not use large wood chips, wood waste from treated lumber, or materials that form a mat, such as long grass or leaves. Dry pine shavings often are sold as bales of animal bedding.

Determining the amount and frequency of additions

Adding bulking material helps control moisture and keep the materials porous and crumbly. If you have a direct-drop pedestal, then bulking additives can be added after each use (1–2 cups). If however you have a

Fig. 7.8: *The Phoenix system for the Bullitt Center in Seattle, Washington, considered to be the greenest commercial building in the world; it was fully certified in the Living Building Challenge.*

toilet that is not suited to having shavings put into it (e.g. micro-flush, foam-flush, vacuum flush) then you will want to add bulking materials through the back door at least every 500 uses.

To determine the quantity of what is added, figure on 4–8 L (1–2 gal) of bulking material for every 100 uses; therefore, if servicing at 500 uses, you would add 5–10 gallons. The more frequent addition of bulking material will reduce mixing requirements. With some semi-public and residential systems, a bin of bulking material can be available to users for "flushing" (Nelson and Kerfoot, 2006).

Leachate can be handled either via gravity drainage or pumping.

Worm composting has become a standard practice for users of the Phoenix; it results in greatly reduced volumes. Due to the fine texture of worm castings, the drainage screen becomes less porous over time; therefore, when worms are going to be used, the screen system is altered (more coarse) to account for the impact to drainage.

Pedestal

These systems can accept a variety of flush applications to match personal preferences. For gravity systems, foam-flush is used; and for systems that require horizontal runs that have too shallow a slope for gravity, vacuum pumps are used.

In cases where the vacuum toilet is used, the collection chamber gets more moisture, which happens to make a happy worm habitat; therefore, worms are commonly added to designs that have such toilets. With the wetter environment, the manufacturer also recommends their tallest model to provide the most effective pile leveling resulting from the extra bank of mixing tines.

Cleaning

Soap and water, vinegar, or hydrogen peroxide can be used; but do not use bleach-based disinfectants (true for any compost toilet).

Servicing

- Additional moisture is applied if the pile is too dry (under 40% moisture content).
- Additional bulking agent is added if the pile is too wet (over 75% moisture content), to ensure balanced C:N ratios, and avoid compaction.
- Turn and mix every 500 uses (average home of four would mean turning every 25 to 30 days).
- Upper pile may need to be raked and evened at the same time as turning the upper tines for mixing.

Insect control

- Screened intake and outlet vents are a requirement, as is the ability to remove them for easy yearly cleaning to ensure proper air ventilation flows.
- For foam-flush or Jet vacuum pedestals, no other controls are required.
- For direct-drop gravity systems, the addition of a flytrap is highly recommended, as pictured in Figure 7.6.

Power

Power is required for the fan; the off-grid DC version is 5 watts (consumes 120 watt hours per day). In cases where a leachate pump is required and a choice is made for an electric one over a hand pump, there will be more power needed.

 Advantages:

- This system overcomes the issues of compaction that are more common with other large container systems.
- Flexibility exists to incorporate jet vacuum systems into an existing house due to various flush options.
- The companies in the U.S. and Canada are very service oriented and easy to connect with.
- The documentation is easily accessed, clear, and informative — covering all aspect that would be required for dealing with inspectors, including a maintenance log.

 Disadvantages:

- Cost: $$$$$ ($7,000–$20,000). Price varies dependent on size, flushing, and conveyance options.

Other Manufactured Continuous Systems

There are several other models of manufactured continuous systems available. They do not all function as described, and therefore we cannot stress enough how important it is that you research customer reviews and read the user manual's maintenance schedules to know what you are getting yourself into. Reading the troubleshooting section is also enlightening. If reading such materials seems overwhelming, complicated, too involved, and if the parts lists seem endless, perhaps that system is not your best choice.

Worm (Vermi) Composting

Worms (the good kind) are not new to compost toilets. Before we consider the toilet systems that are well suited for worm composting, we should wiggle our way into the science of worm digestion.

Worms most commonly used in vermicomposting are *Eisenia fetida,* commonly known as *red wigglers, compost worms, manure worms,* or *trout worms.* They can withstand a wide range of conditions that would kill most other worm species. They survive between 0°C–35°C (32°F–95°F), but they thrive between 18°C–27°C (65°F–80°F), maturing and breeding rapidly. (Hill and Baldwin, 2012; Hill, Baldwin, and Lalander, 2013; Worm Farm Facts, "Red Worms").

Worms enhance the decomposition process greatly, and employing them results in quicker volume reduction — which translates into smaller-sized chambers. The inclusion of worms in moldering and continuous systems greatly extends the periods between servicing, thus allowing systems to service larger numbers of users without requiring larger units.

Nitrogen

Composts and manures have various levels of nitrogen that require conversion. There are three main groupings of nitrogen:

1. Molecular nitrogen, present in the atmosphere as a gas.
2. Organic forms, found in the building blocks that make up proteins (amino acids) and DNA (nucleotide bases).
3. Mineral nitrogen (the form we are most interested in), which can be further sub-categorized as ammoniacal nitrogen (ammonium), nitrate, and nitrite.

Ammoniacal nitrogen (urea and ammonium) can be a toxic pollutant in excess, causing cell damage and death; in the form of *nitrate,* nitrogen is tolerated much better

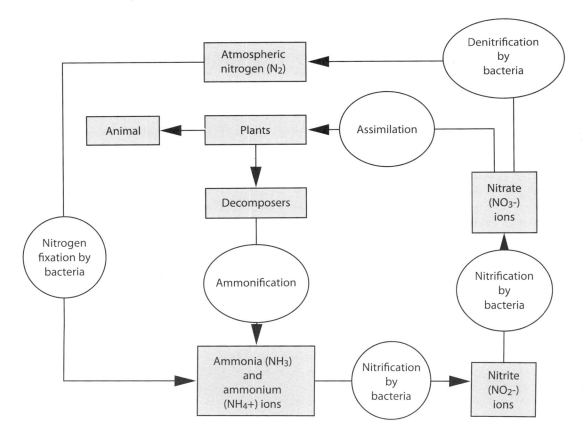

Fig. 7.9: *Schematic representation of the nitrogen cycle.*

Illustration credit: Gord Baird (adapted from Lenntech.com)

by plants and animals. *Nitrite* is a form that exists during the conversion process of ammoniacal nitrogen to nitrate nitrogen (see Figure 7.9).

Nitrogen and worms

What is the link between mineral nitrogen and worms? Bacteria can consume the damaging ammonium and turn it into a form useful for plants. This is the process that fuels a thermophilic compost pile. However, worms carry large colonies of bacteria in their gut, and their gut bacteria do the same conversion. But worms (and their guts) thrive at cooler temperatures. Therefore, with the help of worms, lots of nasty nitrogen is consumed and converted into safe forms for plants without relying on a thermophilic process. You might say that's cool, but really, it's just mesophilic. Groan....

The point where the nitrogen cycles are balanced is when the materials are considered nontoxic to the soil; at that point, the compost is considered mature/cured. This maturation/curing pathway, introduced in Chapter 2, relies on the conversion of this nitrogen, hence the use of Solvita's Compost Maturity Index test (discussed more thoroughly in Chapter 9).

When worms are used in a system, nitrogen conversion can happen incredibly quickly; conventional maturation time can be reduced from 6 months to 8–9 weeks (Hill and Baldwin, 2012; Wichuk and McCartney, 2010). However, it is important to note that *sanitization* is not sped up. Some sanitization aspects are enhanced (there is increased *E. coli* death, but no research has demonstrated that Ascaris ova [eggs] are impacted). For this reason, the rule is to always treat for the worst-case scenario, which means a further sanitization process is required if

you want to surface-discharge materials to the soils. If you don't wish to use the sanitized materials in surface soils, then burial at a minimum of 15 cm (6 in) is acceptable.

What non-thermophilic systems are worms used in?

- Carousel
- Sealed moldering system
- Aquatron
- Ecodomeo
- Phoenix

How are worms introduced?

- As few as six worms introduced into a system will inoculate it.
- 1–2 lbs of worms with 8 cm (3 in) of damp sawdust or leaf mulch is ideal.
- In unsealed systems, worms will find their way to all containers.
- In sealed systems, transferring some worms from one bin to another will work.

Despite dramatically reduced volumes (carousels have seen reduction of volumes up to 90%), initial sizing and servicing frequency should be calculated as if no worms were being used. Only after time and observation should servicing schedules be updated to match actual experience.

Worms have lots of benefits, but worms also come with implications for drainage.

Leachate drains can become clogged with *worm balls* (a highly technical term describing fist-sized clusters of living worms); therefore, design drains to be 1¼" diameter or larger.

Worm castings (worm poop) are very fine particulates; like all excrement, they can clog standard screens. For this reason, coarser prescreens may need to be incorporated in front of the finer standard screen to limit the finer screens from getting plugged up.

Chapter 8

Fluid Management

$\mathbf{F}$LUID MANAGEMENT includes leachate handling for continuous and moldering systems; it also includes urine diversion in any form for any system (either separated at the seat or into a urinal). Systems will fail if excess fluids are allowed to accumulate. Systems that encourage biological processes with thermophilic conditions consume and evaporate large amounts of moisture — but this only occurs in the compost pile, not in the house. Thus, we cannot rely on this biological benefit in collection containers;

we must rely on other methods to avoid the buildup of problematic excess fluids.

There are five ways to address excess moisture as illustrated in Figure 8.1:

1. Urine diversion at the pedestal or waterless urinal (source separated)
2. Fluid separation inside the conveyance area, creating separated wastewater (as seen in the Ecodomeo in Figure 6.11)
3. Leachate drainage of the fluids that filter through the contents of the collection chamber (or compost pile)

Fig. 8.1: *Moisture can be handled by source-separating seats, leachate drains, fans, and heating elements.*

ILLUSTRATION CREDIT: GORD BAIRD

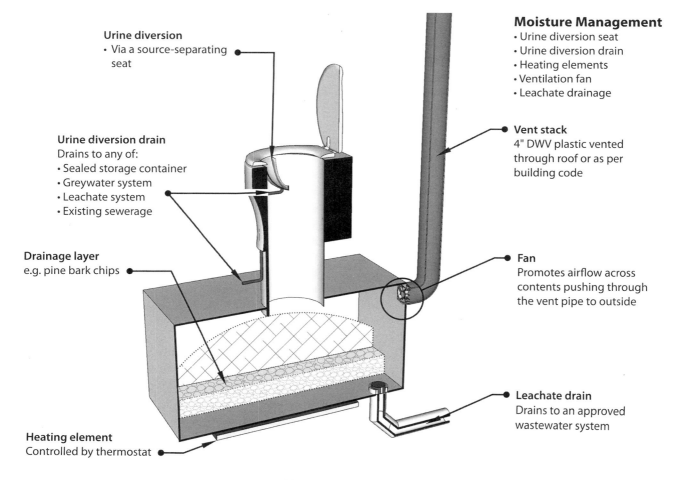

Urine diversion
• Via a source-separating seat

Urine diversion drain
Drains to any of:
• Sealed storage container
• Greywater system
• Leachate system
• Existing sewerage

Drainage layer
e.g. pine bark chips

Heating element
Controlled by thermostat

Moisture Management
• Urine diversion seat
• Urine diversion drain
• Heating elements
• Ventilation fan
• Leachate drainage

Vent stack
4" DWV plastic vented through roof or as per building code

Fan
Promotes airflow across contents pushing through the vent pipe to outside

Leachate drain
Drains to an approved wastewater system

4. Evaporation via air circulation (fans)
5. Evaporation via heating elements

We will start by looking at options for urine, then look at leachate.

Urine Management

Urine management means you:

- get rid of it (send it to a sewerage system)
- get rid of it and feel good about it being somewhat useful (send it to greywater irrigation)
- collect it and use it (fertilizer)

Fig. 8.2: *Volume of urine excreted per year per person.*

ILLUSTRATION CREDIT: GORD BAIRD

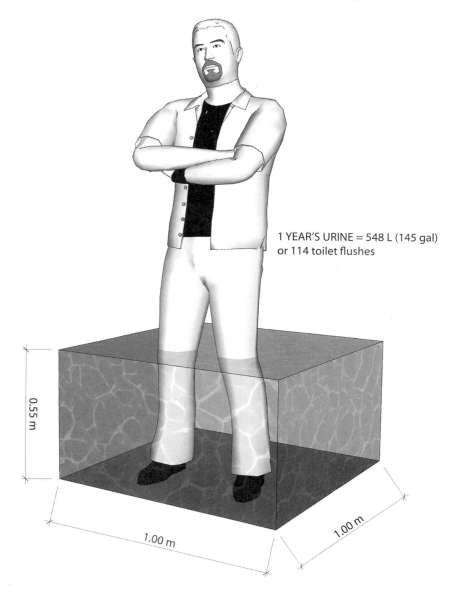

1 YEAR'S URINE = 548 L (145 gal) or 114 toilet flushes

0.55 m

1.00 m

1.00 m

Collecting, storing, diluting, and using urine is a bonus for some, but an inconvenience for others. Combining urine with greywater and dispersing it into the landscape avoids having to manage "another" collection system while allowing one to meet their desire to return nutrients to the landscape. For those who just want it gone, a small urine dispersal or leachate system may be the best solution. In the case of an aging septic system, it is not uncommon for just urine or leachate to be directed into the septic, while solids and other light greywater flows are handled through different means. Before we plan what to do with the urine, we'll pee(k) at it a little closer.

First and foremost, don't assume urine is sterile. The quality of one's urine is directly related to a person's health and lifestyle. Though a healthy individual's urine is sterile in the bladder, bacteria and viruses can be picked up in the urinary tract as it passes through. Additionally, other items may be present in the urine. For example, elderly populations tend to have higher levels of pharmaceuticals in their urine; females on contraceptives have higher levels of hormones; and all of us have heavy metals and persistent organic pollutants (POPs) derived from our environment (von Münch and Winker, 2011). Our good judgement is required when handling urine.

The average person excretes 1.5 L per day (50 oz), 45.6 L (12 gal) per month, or 548 L (145 gal) per year (illustrated in Figure 8.2). That's lot, and if we recall the wine glass in Figure 2.4, we remember there are a lot of solids (or minerals) in that urine.

Urine accounts for 80% of the nitrogen excreted from the human body. The nitrogen in our urine is in the form of urea, which is both useful and problematic. Although it leaves us as urea, once stored, it converts to ammonia, which is ammonium plus hydrocarbonate. This conversion process results in a higher pH (near 9), which enables undiluted urine to naturally sanitize itself over time. Very beneficial! (von Münch and Winker, 2011). The downside is that the higher pH also results in the creation of drain-clogging crystals of *struvite* and calcium phosphate, which precipitate out of the urine (Larsen and Lienert, 2007). Not beneficial.

So back to the fluid dynamics: where does urine originate in our toilet design?

Waterless urinals

Waterless urinals are now commonly employed in public facilities. The first waterless urinals were introduced in 1894, in Austria. But the technology garnered new interest in the early 1990s, when water conservation and the topic of ecosan (ecological sanitation) became hot topics. A urinal functions by allowing urine to flow to a drain that is equipped with a device to stop odor from escaping up and out the drainage system. These odor traps, or seals, include:

Blocking fluid (liquid sealant) — The blocking fluid, or liquid seal trap, incorporates a fluid that is lighter than water and thus floats on water. This fluid resides inside a U-bend or P(ee)-trap. As urine enters the trap, the urine, heavier than the fluid, falls through and drains past the trap into a drain, leaving the fluid behind to act as an odor barrier. This sealant can be oil or aliphatic alcohol (like ethanol) and is biodegradable. Sealing fluids need to be replaced over time. See Figure 8.3.

Fig. 8.3: *Example of a blocking fluid valve used in urinals and urine-diversion system.*

ILLUSTRATION CREDIT: GORD BAIRD

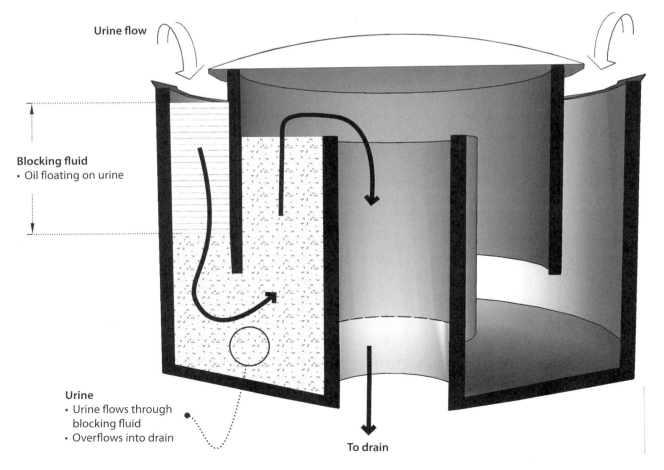

Urine flow

Blocking fluid
• Oil floating on urine

Urine
• Urine flows through blocking fluid
• Overflows into drain

To drain

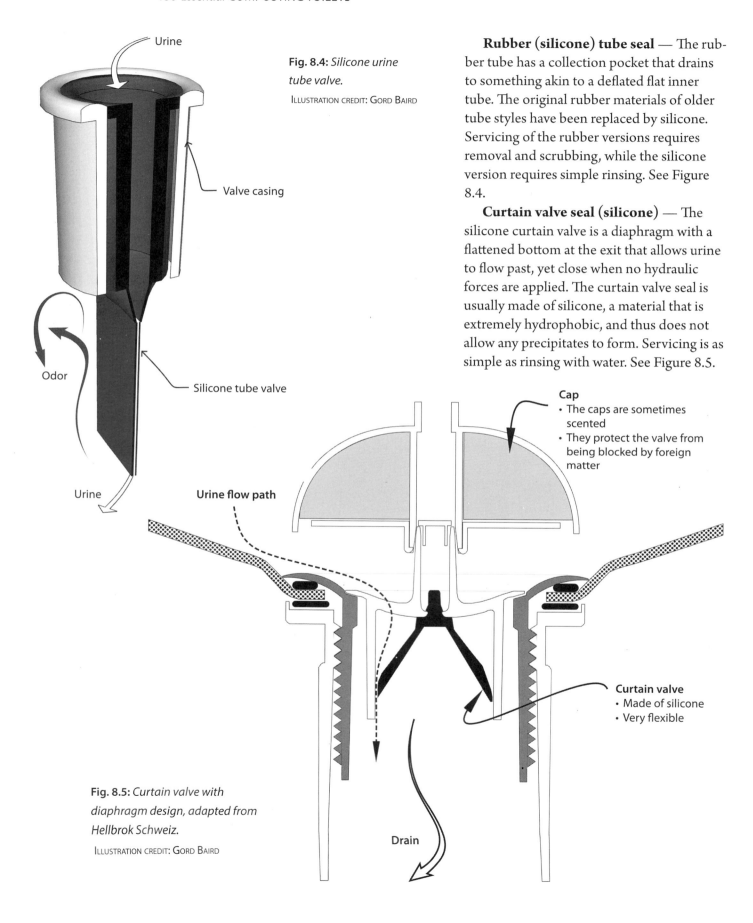

Urine

Valve casing

Odor

Silicone tube valve

Urine

Fig. 8.4: *Silicone urine tube valve.*
Illustration credit: Gord Baird

Rubber (silicone) tube seal — The rubber tube has a collection pocket that drains to something akin to a deflated flat inner tube. The original rubber materials of older tube styles have been replaced by silicone. Servicing of the rubber versions requires removal and scrubbing, while the silicone version requires simple rinsing. See Figure 8.4.

Curtain valve seal (silicone) — The silicone curtain valve is a diaphragm with a flattened bottom at the exit that allows urine to flow past, yet close when no hydraulic forces are applied. The curtain valve seal is usually made of silicone, a material that is extremely hydrophobic, and thus does not allow any precipitates to form. Servicing is as simple as rinsing with water. See Figure 8.5.

Cap
• The caps are sometimes scented
• They protect the valve from being blocked by foreign matter

Urine flow path

Curtain valve
• Made of silicone
• Very flexible

Drain

Fig. 8.5: *Curtain valve with diaphragm design, adapted from Hellbrok Schweiz.*
Illustration credit: Gord Baird

SSUD Seats

Toilets that rely on source-separating urine diversion (SSUD) incorporate source-separating seats (see Figures 3.8 and 5.11), which allow solids to move to a chamber and urine to move elsewhere.

Manufacturers often provide the option of choosing either a source-separating seat or a regular seat for each of their units (therefore, before you purchase, you'll already want to have decided which you want). SSUD seats can also be purchased separately for site-built, DIY systems.

An online search for urine-diverting seats will offer many alternatives — though you may want to avoid such arrangements as the funnel tied to the toilet seat with bailing twine — a method that will not score any points with an inspector.

In Chapter 3 we discussed how urine diversion can be a tough sell because source-separating seats are still unusual to find in North American (Larsen and Lienert, 2007). Integral to the functioning of SSUD compost toilets is behavioral change; therefore, they work best when used in homes where users are keen to develop the "correct" pee and poo habits. Public use is almost always a concern due to male habits; the result is daily maintenance and supervision (for the toilets, that is).

Key Design Elements for Source-Separated Urine Diversion Systems

- No parts of a diversion system should be metal (Ralston, 2016, p. 85).
- Use smooth-walled plastic (hydrophobic) materials for piping (HDPE, ABS, or PVC; kink-proof silicone tubing may be the very best choice, though).
- Avoid drainage design that slows flow because this increases retention time and thus precipitation (so, avoid siphons, horizontal piping, P-traps, etc.).
- Diversion plumbing should be a minimum of 1¼" plastic piping.
- Plumbing slope should be a minimum of 4% (exceeding most plumbing codes, which stipulate 2%).
- Urine-diversion piping should be as short as possible; avoid lengths longer than 2 m (6 ft).
- Allow for flushing with soft and/or warm water.
- Employ design features that allow unimpeded flow but control for odor (discussed later in this chapter under "Liquid Urine Storage").

Maintenance of SSUD systems

Good housekeeping — Reduce the inclusion of anything that provides additional surfaces for struvite to precipitate out onto, such as hair and/or the odd bit of toilet paper that may enter the urine diversion tubing.

Good plumbing — Horizontal plumbing must be avoided, as should plumbing elbows of 90°; instead, use long-sweep elbows, or two 45° elbows.

Good maintenance — Urine odor seals (often used in waterless urinals) need periodic servicing or replacing. Cleaning of components may include mechanical cleaning with caustic soda (2 parts water, 1 part caustic soda), vinegar, or strong acetic acid (24%). Urine-diversion plumbing should be flushed at least once per year, a fact that should be explicitly stated in any maintenance schedule documentation.

SSUD Use Options

As mentioned at the beginning of the chapter, diverted urine can be sent in many directions. If you just want to dispose of it, then you can treat it the same as you would a leachate, which is discussed later in this chapter.

If you wish to use urine's nitrogen and micronutrients as a fertilizer, fresh urine can either be diluted and used directly in the garden or incorporated into subsurface greywater dispersal systems.

Greywater

If one needs urine to go "away," but still desires nutrients to be utilized in the landscape, then there are options for directing the urine to greywater reuse systems.

Greywater is water from general household sources such as sinks, bathtubs, showers, laundry, and kitchen sinks, but not toilet water or leachate containing feces. All greywater sources are not equal though, as different sources have different characteristics.

Greywater is classified from light grey to dark grey depending on the potential level of organics, bacteria, and suspended solids. The table in Figure 8.6 succinctly defines greywater categories. For example, kitchen water is considered dark grey because it potentially has more solids, food particles, and pathogen loading than the light greywater of showers.

In greywater reuse there are three basic rules:

1. We do not apply to food crops, except perennial woody tree crops (dicots).
2. We do not store it for more than 24 hours unless it gets aerated (otherwise it becomes anaerobic and stinky).
3. We do not apply it above ground, thereby removing opportunities for pets, kids, or others to unknowingly come into contact with it.

These three rules meet the needs of the F Diagram, first seen in Chapter 2 (Figure 2.15), which is important because greywater (with urine or no urine) is potentially a pathogenic water source.

The various methods for dealing with greywater merit their own book; for our purposes, we'll look at the simplest, easiest, cheapest, and most common method of distribution: the branch drain method, shown in Figure 8.7.

Abbreviation	Category	Description
Laundry greywater		Washing machine & laundry sink
LGW	Light greywater	Bathtubs, showers, bathroom sinks & laundry
VLGW	Very light greywater	Bathtubs, showers, bathroom sinks
DGW	Dark greywater	Dishwasher, kitchen sinks, bar sinks, mop sinks
CG	Combined greywater	All residential wastewater except flush toilets & urinals
Blackwater		Flush toilets and urinals
Brownwater		Flush toilets with urine diverted
Water separated from blackwater		Water separated from blackwater or brownwater
Urine		Urinals and/or diverted urine
Leachate		Fluids collected from compost processors
Combined sewage		Blackwater and greywater

Fig. 8.6: *Source-separated wastewater streams.*
Credit: Ralston, 2016

In this system, urine enters the greywater stream and becomes naturally diluted with the other household flows. This has the benefit of flushing urine precipitants out of the pipework. Urine entering the stream in this manner does not impact the existing drain-waste-vent (DWV) system, as it can vent out the stacks and is downstream of all P-traps. The sizing of the mulch basins (Figure 8.7) will be determined by flow volume, the soils, and the type of fluids. The next section, on designing a leachate system, has details, including calculations and design criteria which can be applied to designing a greywater mulch basin.

The patchwork of greywater guidelines and regulations across North America is slowly evolving. Regulatory authorities are

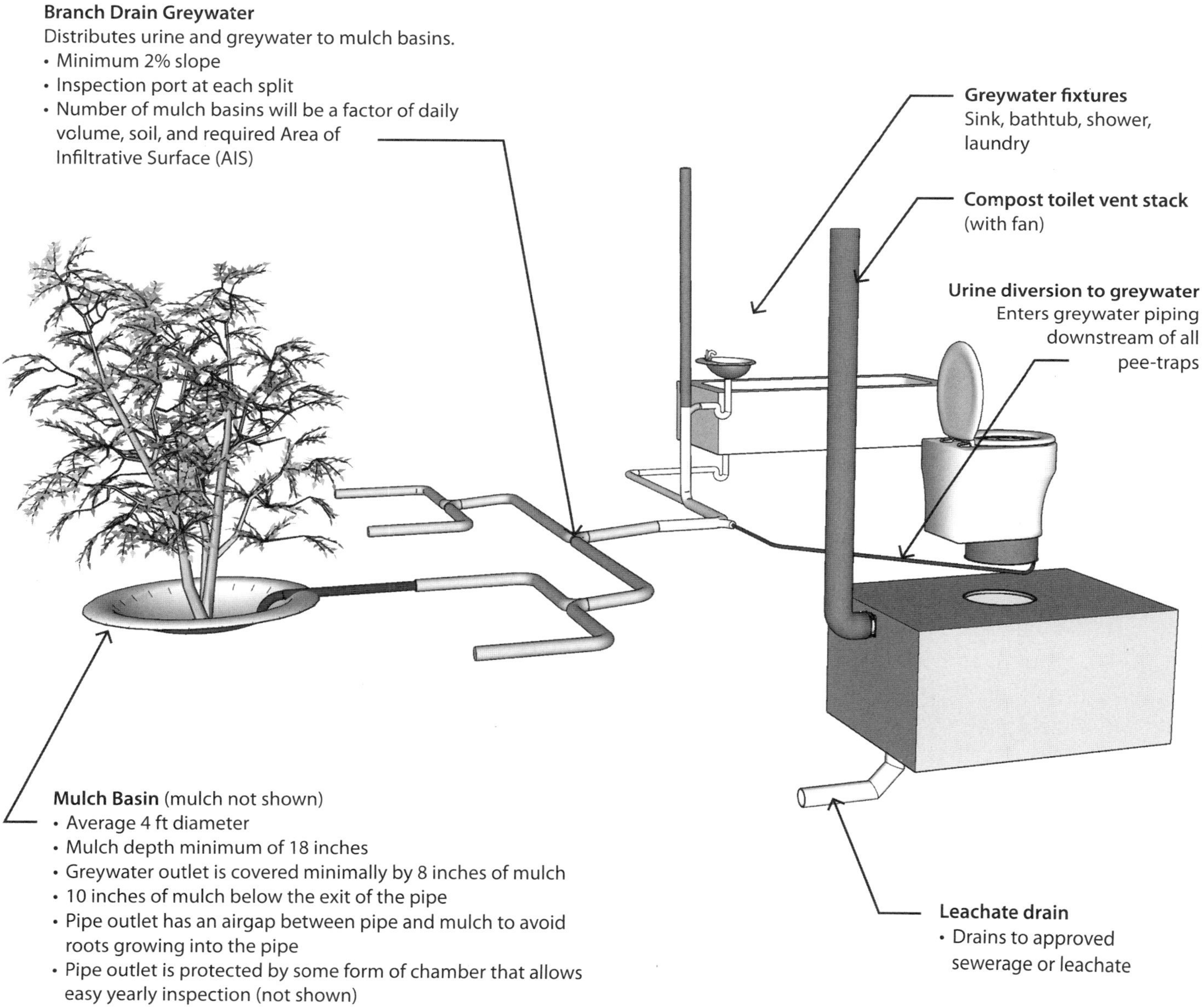

Fig. 8.7: *Urine diversion to greywater (branch drain).*
ILLUSTRATION CREDIT: GORD BAIRD

Branch Drain Greywater
Distributes urine and greywater to mulch basins.
• Minimum 2% slope
• Inspection port at each split
• Number of mulch basins will be a factor of daily volume, soil, and required Area of Infiltrative Surface (AIS)

Greywater fixtures
Sink, bathtub, shower, laundry

Compost toilet vent stack (with fan)

Urine diversion to greywater
Enters greywater piping downstream of all pee-traps

Mulch Basin (mulch not shown)
• Average 4 ft diameter
• Mulch depth minimum of 18 inches
• Greywater outlet is covered minimally by 8 inches of mulch
• 10 inches of mulch below the exit of the pipe
• Pipe outlet has an airgap between pipe and mulch to avoid roots growing into the pipe
• Pipe outlet is protected by some form of chamber that allows easy yearly inspection (not shown)

Leachate drain
• Drains to approved sewerage or leachate

exploring ways to reduce load volumes on aging infrastructure and respond to drought and water scarcity. For those who have always had an ecological motivation for greywater use/reuse, it is refreshing to see cities, counties, states, and provinces finding value in adaptation and mitigation of the risks and impacts associated with changing precipitation patterns due to climate change.

If greywater is not to be included in your water system design, then urine storage systems are your next option.

Liquid Urine Storage

Urine storage can be small scale (daily use) or scaled up if you intend to use it as a liquid fertilizer. As you will see, it can also be stored and dried to create a dry fertilizer. Let's look at the liquid form first.

Small (daily) collection

The simplest form of collection is the daily collection into a small "pee" container that is applied to the landscape every 1–2 days (Nature's Head and Air Head toilets demonstrate this well. See Figure 7.1). This approach is functional at the small-household scale because the pee can be used without being sanitized, due to the very low risk factors. In this case, urine is diluted 10:1 with water before application to plants, or it can be applied full strength directly to the compost pile. The benefit of applying it to the compost pile is that it will support the thermophilic processes, helping you meet your curing/sanitizing goals. For small-scale immediate-use settings, calculations of storage container "sizing" is not required (WHO, 2006).

Large (monthly to yearly) collection

If it is impractical to provide frequent servicing and/or urine cannot immediately be used, then larger storage is an option. Considerations that would help decide this are:

- Toilets that receive large fluctuating peak use (such as when hosting a party) and there is an inability to adjust servicing to match the influx.
- Climates in which the soil is below 10°C (50°F) for weeks on end (meaning the soil organisms will not provide any biological activity to promote degradation).
- Climates in which temperatures fall below −5°C (23°F) — because urine freezes.

A final consideration relates to the future. Despite the present user being willing to perform the daily dispersal of urine, a future owner may wish to have a storage system or a drain for a "get rid of it" system. Our advice would be to install the storage system and design for future owners who may wish to divert. Leave your options open.

Key design elements for liquid urine storage

When liquid storage is required, urine should be stored undiluted for a sufficient time to allow the urea's conversion to ammonia to sanitize the stored materials. This aging time should be 1–3 months for residential systems, or 6 months for public systems (WHO, 2006, page 39; Spuhler, 2018).

Liquid urine container sizing

When size matters: Container sizing for *stationary* tanks: (See table right)

Calculations: *What Container Size for the urine (CSu) do I need?*

$$CSu = (O \times U \times T \times f)\,(Sf)$$

$$CSu = (4 \times 1.5\,L \times 90 \times 1)\,(110\%)$$

$$CSu = 594\,L$$

Therefore, a four-person home will require storage enough to hold 594 L (157 gal) over 90 days.

When size doesn't matter: Container sizing for *moveable* 20 L (5.3 gal) containers:

Sizing is not required when movable 20 L storage containers are used; when full, 20 L containers can be replaced by a new container, the full one is set aside and labeled with a date. This unlimited flexibility works on the same concept of flexible servicing found with commode batch compost toilet systems.

- Using 20 L containers work well for those systems where homeowners may wish to use their urine immediately in the growing months, but store it in the winter.
- A sufficient number of storage containers are required: one for the collection period and enough for containing urine through the entire aging period (Chapter 4 covers calculations).
- Containers need to be sealed to keep ammonia gas in and critters out.
- Inlet pipes (carrying urine) into the container should extend down to the base, filling the tank from the bottom and thus acting as an air lock.

O = # of occupants/users	4
U = Urine production/person/day	1.5 L
T = desired storage time	90 days
F = fraction of time a user will be on premise in a day	100%, or 1.0
Sf = Safety factor*	110%, or 1.1

(*Container should exceed the required storage by a safety factor of 10% represented as 110%.)

- Minimal but proper venting of the inlet pipe is required for pressure equalization.
- If the tank is pumped out, then the container requires an air inlet that allows air inflow equivalent to the pumping of fluid to stop the tank from imploding due to the creation of a vacuum.
- For stationary tanks, an overflow drainage port from the tank is suggested, where the overflow drains to an appropriate system as noted below.

An owner's manual for a urine diversion system with storage would include:

- Number of users per day for the design.
- Size of collection tank.
- Procedure for pumping or container swapping.
- Instructions for where to store full containers.
- Instructions for storage time required before use (must be stored three months before using for it to self-sanitize).
- A labeling scheme for containers in rotation. (Ensure that when a full container is swapped out and set to sanitize that the date of "DO NOT USE BEFORE ____ " is attached.)
- Cleaning procedures for tubes, tanks, and drains.

Ventilation options for urine diversion seat

Odors (ammonia) will migrate up out of the urine diversion plumbing despite the outlet being submersed into the storage tank. The three options to control for this (other than urine traps) are shown in Figure 8.8:

- Option 1: Storage vent tube connected to the main mechanically vented exhaust stack.
- Option 2: Storage vent tube connected back into the urine diversion plumbing in a manner that does not allow for urine to gain access to the tube.
- Option 3: Equalization hole in the urine diversion tube located as close to the diversion piping entrance to tank as feasible.

If the urine diversion drain acts as a vent (options 2 and 3), either the toilet pedestal has to be ventilated, or, for systems with privacy flaps or valves where suction from the main solids collection chamber is blocked, you can vent the urine diversion pipe to the main mechanically vented stack (the dotted line in Figure 8.8 depicts optional venting).

Urine storage overflow

Accidents happen. Design for it. The illustration in Figure 8.9 shows options for providing overflow protection.

- Option 1 in the diagram uses a P-trap or a sewer-grade swing check valve to trap gasses inside. Either of the two options (P-trap

Fig. 8.8: *Urine storage tank ventilation options include active and passive methods.*
ILLUSTRATION CREDIT: GORD BAIRD

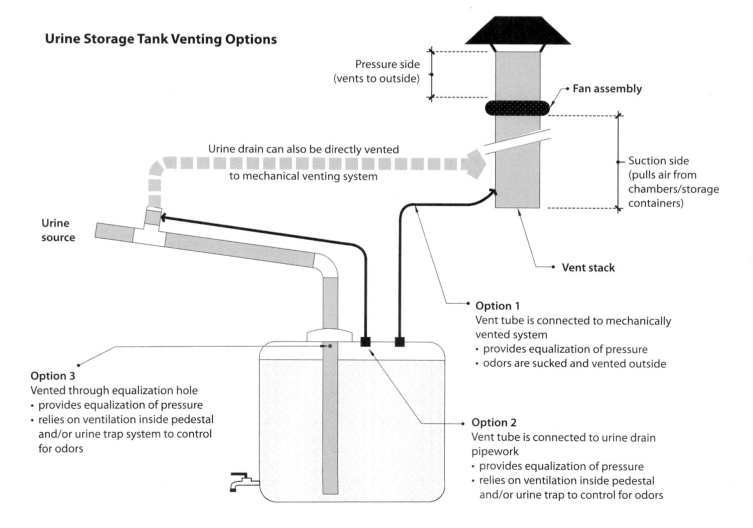

Urine Storage Tank Venting Options

Pressure side (vents to outside)

Fan assembly

Urine drain can also be directly vented to mechanical venting system

Suction side (pulls air from chambers/storage containers)

Urine source

Vent stack

Option 3
Vented through equalization hole
- provides equalization of pressure
- relies on ventilation inside pedestal and/or urine trap system to control for odors

Option 1
Vent tube is connected to mechanically vented system
- provides equalization of pressure
- odors are sucked and vented outside

Option 2
Vent tube is connected to urine drain pipework
- provides equalization of pressure
- relies on ventilation inside pedestal and/or urine trap to control for odors

or check valve) need to be serviceable with a clean out.

- Option 2 in Figure 8.9 provides access to inspect or insert a pump and pump out the tank while also providing the required overflow and seal required to trap ammonia gas.

To address the sludge that will accumulate in the bottom of the tank, you will need to consider a method to give the ability to stir up the settled solids and make a slurry for pumping or draining; this will be part of the maintenance.

Application of stored liquid urine

Application of the daily collected volumes is fairly straightforward, but when larger volumes are collected, then it is a different scope of management. Consider: if you have collected and stored 500 L of fluid and you want to apply it to the landscape, what does this look like? If you dilute it at a rate of 10:1 it looks like 5,000 L (1,321 gal).

People who consider urine for fertilizer land-use application usually have a means of transferring it by truck and tank to the fields or orchards and a sprayer system to apply it with. If you would like more information on using stored self-sanitized urine, we recommend you explore the resources listed at the end of this chapter, as this topic has moved into the agricultural aspects and is beyond the scope of this book.

Fig. 8.9: *Urine storage tank overflow options.*

ILLUSTRATION CREDIT: GORD BAIRD

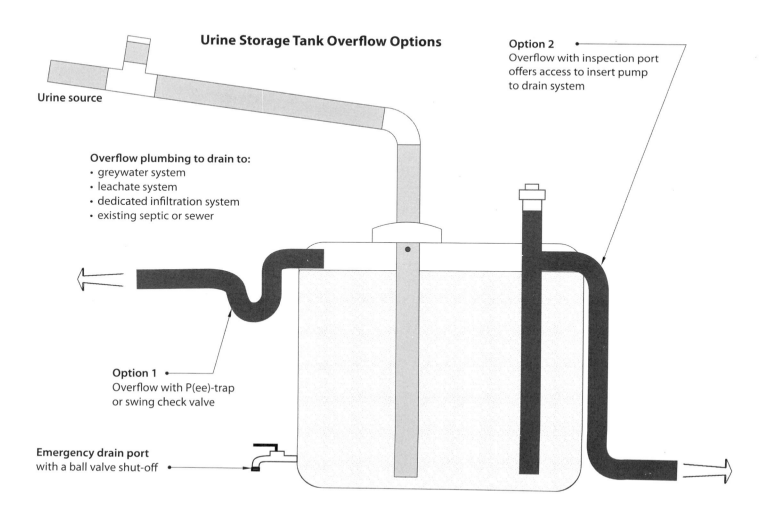

Urine Storage Tank Overflow Options

Urine source

Option 2
Overflow with inspection port offers access to insert pump to drain system

Overflow plumbing to drain to:
- greywater system
- leachate system
- dedicated infiltration system
- existing septic or sewer

Option 1
Overflow with P(ee)-trap or swing check valve

Emergency drain port
with a ball valve shut-off

Dry Storage

Fluid storage for fertilizer use can be cumbersome if you are not properly set up. Here is another option that is being investigated, and one we are experimenting with at home. Collecting urine into wood ash and drying it maintains the fertilizer value, yet avoids the bulky and heavy aspects tied to liquid. The dried fertilizer can be spread by hand.

Research in 2017 looked at the possibility of diverting urine into wood ash at a rate of 20 L per kg of ash. Wood ash works by preventing urea's conversion into ammonia because the high pH (>12) inhibits enzymatic activity, keeping urea in its unaltered form, hence less nitrogen loss. The concept arose from the need to address issues of liquid storage for higher volume (higher traffic) toilets, the transportation of heavy containers of fluid to the area of land application, and the equipment (and maintenance of that equipment) used in land application. The dry product is lighter to transport, easier to store, and requires less equipment to apply. At the time of writing, a system had not yet been finalized, but the conclusions of the research do offer design considerations, including adding urine to the ash rather than adding ash after each pee, mechanical mixing to distribute the urine and assist evaporation, and the addition of heat to help evaporation (Senecal and Vinnerås, 2017).

We began experimenting with ash use in our nighttime urine-only commode and found it to eliminate any urea odors, even after three weeks of continual use. Prior to using the wood ash, the ensuite commode had to be changed weekly. We prepare the ensuite bucket with 2.5 cm (1 in) of shavings on the bottom and 10 cm (4 in) of ash topped with another 2.5 cm of shavings; then, a handful of shavings is thrown in after each use. See Figure 5.2. Where would you find suitable wood ash? Woodstoves, fireplaces, and fire pits are a good start. Farm animal feed stores may have some, as wood ash is used to treat for mites.

Key design elements for dry storage

- A storage container should be sized to accept 20 L (5 gallons) of urine per kg (2.2 lbs) of wood ash.
- Adding heat to bring temperatures to 65°C (149°F) will allow heavier loading rates of urine (by evaporating excess moisture more quickly).
- An ash bed is preferable to adding ash after each dose of urine.
- Stirring/turning the ash bed is useful for homogenizing the urine/ash mixture.

Urine and Fluid Separation

What is leachate? Systems that collect feces and urine together and then *separate* the fluids before they enter the collection chamber differ from "diversion," in that the fluids have come into contact with feces and will contain contaminants found in the feces. This includes the systems that use micro-flush and jet vacuum flush, which add extra fluids that then require removal and disposal. All these fluids must be treated as *leachate* — an equivalent to what we may call "combined sewage" or "water separated from blackwater" — and disposed of through a designated wastewater system or other approved sewerage system. The essential point here is that separated waters are vastly different from diverted urine.

Only waters that are treated and deemed sanitary — like those that come out of the Aquatron units discussed in Chapter 6 (in which fluids are UV treated and phosphorus removed) — can go into the greywater.

Leachate Drainage Systems

Many compost toilet designs will require a drainage system for at least the leachate from the compost processor. The following discussion focuses on leachate, but the methods and design considerations are applicable across the whole range of effluents listed in Figure 8.6.

Leachate is the fluid draining from the compost processor, and it has all the qualities that would classify it as blackwater, except it is more concentrated with all the nasty things we are trying to control for, because it is not diluted with flush water. For this reason, leachate drainage systems need to be designed like a septic system, with tanks to allow adequate transit times for biological processes to occur before fluid is discharged to subsurface infiltration beds. These beds must be able to handle the volume (hydraulic load). The consequences of not having appropriately sized tanks and adequate infiltration systems to match the daily flows is that the organics, suspended solids, and bacterial residues will quickly plug the infiltration system.

It is critical to think of leachate drainage as a treatment system and not as a simple drain pit or soak-away pit, which is how most manufacturers of compost toilets present them. It is about treating it — not getting it "away"; if not, the "away" comes back to haunt you.

Due to the lower flow volumes, leachate systems can be less expensive than septic systems because they require smaller tanks, less excavation, and less materials. Design costs vary widely though, depending on whether other systems for greywater or combined wastewater are required. The combined costs for a full system could be on par with a full septic system.

This discussion of leachate systems will focus on an understanding of wastewater flows in general; so, in an attempt to explain leachate, you'll be introduced to the broader concepts of sewerage system design. If you plan on using a professional to design your leachate system, then you can skip the topic of sizing. Then again, if you are unsure whether you will hire this out or do it yourself, this information will help you make that choice.

Remember that if you choose a batch commode all-in system (like the Humanure system), you will not have any leachate to deal with. And, if you plan on hooking up your leachate to an existing septic system or an existing sewerage system, you will not need to design a new leachate system. There are a few other situations in which you will not require a leachate system. The following information is intended for those wishing to design for themselves a new system for any combination of leachate, urine diversion, or greywater.

Sizing Leachate Systems

Wastewater and sanitation science is the hinge-pin supporting conventional design practices, regulations, and standards of sewerage systems. Using existing tools and practices of design — and adjusting as needed — allows designers and regulators to more easily understand the sizing of alternate wastewater streams.

Conventional septic system design (combined sewage, inclusive of all household wastewaters) is sized around the concept of the Daily Design Flow (DDF). This is the expected volume a sewerage system is designed to handle. The DDF is influenced by the number of potential full-time users and the composition of the effluent. Both of these factors determine the size of the septic tank and size of the infiltration system into the landscape.

The leachate system is designed like that of a conventional septic system, except there are adjustments made to the DDF volume and the tank sizing to account for both a lower volume of fluids and a higher concentration of suspended solids than conventional combined sewage. Due to these variables, adjustments are made to the standard DDF and tank sizes, resulting in *Adjusted Flow Volume (AFV)* and *Adjusted Tank Sizing (ATS)*; these are calculated using an adjustment factor (a multiplier), which is applied to the standard calculations.

Flow Volumes and Tank Sizing

This is a complicated section. We start by introducing the three main steps in sizing a leachate system. Figures 8.10, 8.11, and 8.12 will aid in visualizing the steps and the processes and show you how to locate needed information on the tables to perform the analysis.

We follow the introduction to these steps with a step-by-step walk through of some examples.

Step 1: Flow Volume (Figure 8.10)

Determine the *Adjusted Flow Volume (AFV)* for the specified effluent:

- Determine occupancy level.

- Calculate the Daily Design Flow (DDF) based on the # of occupants.
- Multiply the DDF by the *flow volume adjustment factor* to determine the AFV.

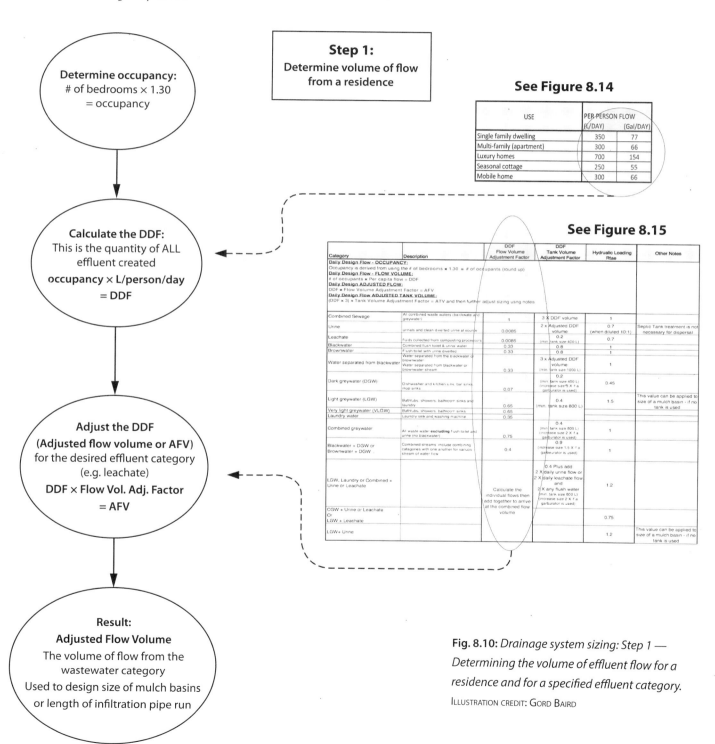

Fig. 8.10: *Drainage system sizing: Step 1 — Determining the volume of effluent flow for a residence and for a specified effluent category.*

Illustration credit: Gord Baird

Step 2: Determine Tank Size (Figure 8.11)

Determine the *Adjusted Tank Size (ATS)* for the specified effluent:

- Determine occupancy level.
- Calculate the DDF based on the # of occupants.

- Calculate tank size for the combined sewage/effluent flows.
- Calculate the ATS by multiplying the DDF tank size by the tank size adjustment factor.

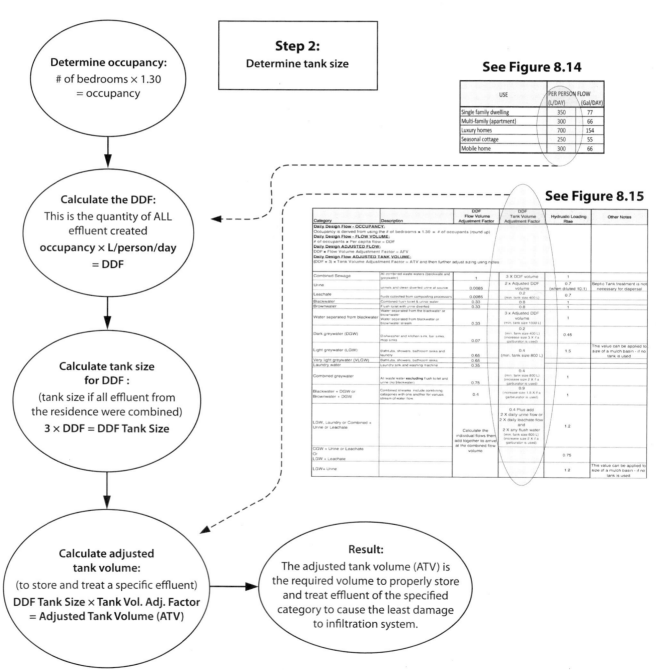

Determine occupancy:
# of bedrooms × 1.30
= occupancy

Step 2:
Determine tank size

See Figure 8.14

USE	PER PERSON FLOW (L/DAY)	(Gal/DAY)
Single family dwelling	350	77
Multi-family (apartment)	300	66
Luxury homes	700	154
Seasonal cottage	250	55
Mobile home	300	66

Calculate the DDF:
This is the quantity of ALL effluent created
occupancy × L/person/day
= DDF

See Figure 8.15

Calculate tank size for DDF :
(tank size if all effluent from the residence were combined)
3 × DDF = DDF Tank Size

Calculate adjusted tank volume:
(to store and treat a specific effluent)
DDF Tank Size × Tank Vol. Adj. Factor
= Adjusted Tank Volume (ATV)

Result:
The adjusted tank volume (ATV) is the required volume to properly store and treat effluent of the specified category to cause the least damage to infiltration system.

Step 3: Infiltration (Figure 8.12)

There are two methods for dispersing fluid into the landscape: apply it over a centralized area, such as a mulch basin; or apply it to a linear area, such as a trench. To determine the area required for infiltration:

- For mulch basins, find the *Area of Infiltration Surface (AIS)*.
- For trenches, use the System Contour Length (SCL) method. See Figure 8.13.

Fig. 8.12:
Drainage system sizing: Step 3 — Determining the volume of effluent flow for a residence and for a specified effluent category.

ILLUSTRATION CREDIT: GORD BAIRD

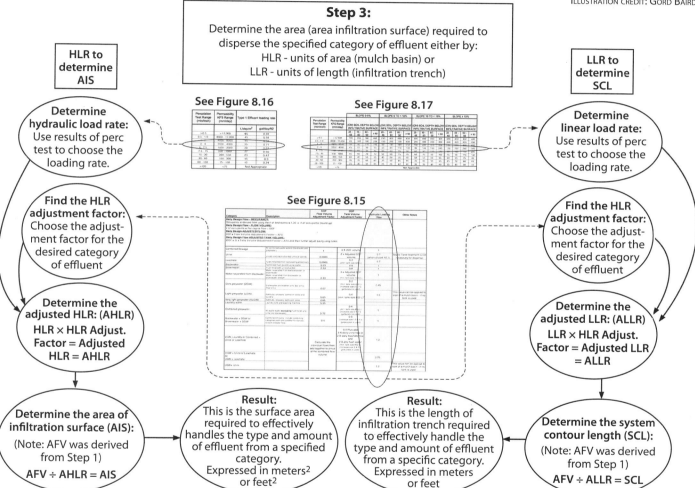

Area of Infiltration Surface Using Hydraulic Loading Rate method (HLR)	System Contour Length Using Linear Load Rate method (LLR)
Used to determine the area required for infiltrating the fluid loads	Used to determine the length of infiltration piping required to disperse the fluid loads
1. Assess soil & perform percolation test	1. Assess soil & perform percolation test
2. Determine hydraulic load rate (HLR)	2. Determine linear load rate (LLR)
3. Determine adjustment factor to apply to HLR	3. Determine adjustment factor to apply to LLR
4. Calculate the adjusted hydraulic load rate (AHLR)	4. Calculate the adjusted linear load rate (ALLR)
5. Calculate the area infiltration surface by dividing the AFV from Step 1 by the AHLR	5. Calculate the system contour length by dividing the AFV from Step 1 by the ALLR

Fig. 8.13: *Leachate Infiltration Methods: One of two methods are used, depending on how one chooses to disperse fluids into the landscape.*

CREDIT: GORD BAIRD

Step 1: Flow Volume (Figure 8.10)

There are several methods for determining the number of potential occupants for a residence other than merely counting them. For example, a 10,000 ft² home with six bedrooms that houses two people today could potentially be home to eight people down the road. If systems are designed around the existing two people, systems will fail if and when usage patterns change to match the building's potential. Jurisdictions usually have their own formulas, which rely on a combination of floor area and bedroom number to determine future potential occupancy.

Using our jurisdiction as example (Victoria, BC), the formula considers the number of bedrooms and the amount of *additional floor area* beyond the *base floor area* allocated to those bedrooms. They apply a calculation wherein this *additional floor area* is divided by 300 to determine the additional people the building structure could support, above and beyond just the bedroom count. Complex? Sure. Need to worry about that? No. The essential point is that design should look down the road at the realistic number of occupants that could potentially reside in a building rather than designing for just present use.

For the purpose of this book and the following examples, we will use 1.3 people/bedroom as our occupancy number.

When we begin to assess Daily Design Flows (DDF), we have to consider both the potential number of occupants and the type of dwelling. The type of residence influences how individuals use water. Those who live in luxury homes use (or perhaps waste) twice as much water as those living in regular single-family dwellings (Figure 8.14). For the purposes of the calculations that follow, we will posit a single-family dwelling that has three bedrooms.

STEP 1a: Determine Occupancy

Using 1.3 people per bedroom, occupancy for a 3-bedroom home would be calculated with the simple formula:

Example:

1.3 people × 3 bedrooms = 3.9

(round up to 4 people)

STEP 1b: Daily Design Flow (DDF)

Once the occupancy is known, you can and then multiply by the *per person flow rate* from Figure 8.14. As noted, we are using a single-family dwelling as our case study.

Example:

4 people × 350 L/day = 1,400 L/day

DDF volume is 1,400 L/day (370 gal)

This DDF is the volume of all potential effluent, combining ALL the potential categories: blackwater, urine, greywater, and leachate. If we want to design a drainage system for leachate alone, we need to adjust the DDF by removing the volumes not related to leachate volume (next step).

STEP 1c: Adjusted Daily Design Flow Volumes (AFV)

To pull that leachate volume out of the mix, we multiply the DDF with a number that

Fig. 8.14: *Per capita daily design flow for residences.* CREDIT: ADAPTED FROM SEWERAGE SYSTEM STANDARD PRACTICE MANUAL V3, TABLE II-9, RALSTON ET AL., 2013 (II-21)

Use	Per Person Flow	
	(L/Day)	(Gal/Day)
Single family dwelling	350	77
Multi-family (apartment)	300	66
Luxury home	700	154
Seasonal cottage	250	55
Mobile home	300	66

represents the percentage of flow that the leachate accounts for. That percentage is our *Adjustment Factor*. When the adjustment factor is applied to the DDF, we receive a flow volume specific to leachate, this is the *Adjusted Flow Volume (AFV)*. These adjustment factors are found in Figure 8.15 in the third column (DDF Flow Volume Adjustment Factor).

The AFV is required when we go to size the dispersal system.

Example:

Adjustment Factor for leachate is 0.0085

AFV = DDF × 0.0085

AFV= 1,400 L/day × 0.0085

AFV = 11.9 L/day (3 gal)

STEP 2: Determine Tank Size (Figure 8.11)

Adjusted tank sizing is required because different fluids have different constituents that impact biological activity. This section walks you through how we determine the appropriately sized tank for the number of people and the type of fluid we are collecting for dispersal.

STEP 2a: Tank sizing for standard daily design flow

Tank sizing starts from the beginning by determining occupancy and figuring out the DDF (step 1a and 1b). Tanks are sized three times the calculated DDF; from this we extrapolate the size of the leachate tank; obviously, it will be smaller than one serving the full DDF. Continuing with the example for the 3-bedroom home with four people, the tank required for a conventional combined sewage system would be 1,400 L multiplied by 3, meaning a tank of 4,200 L.

Step 1a = Occupancy:	3 bedrooms × 1.3 = 3.9 (round to 4)
Step 1b = DDF:	4 people × 350 L/day = 1,400 L/day
DDF Tank Size:	1,400 L × 3 = 4,200 L (1,110 gal)

STEP 2b: Adjusted Tank Size for specified wastewater category

Knowing the DDF tank size now allows us the ability to calculate the Adjusted Tank Size (ATS), the size that best fits the storage and treatment of leachate (in this case). The adjustment factors that we use to adjust the tank size are derived from the characteristics of the effluent (organics, nutrients, dilution) and safety margins. For this reason the *tank volume adjustment factor* is different than the *flow volume adjustment factor* that will be used in step 1c.

Working from the DDF Tank Size (step 2a), we adjust by applying the adjustment factor taken from the table in Figure 8.15, column 4 (DDF Tank Volume Adjustment Factor) for the specified category of wastewater we are working with.

Example:

Tank size = 4,200 L

Tank volume adjustment factor for leachate is 0.2

4,200 L × 0.2 = 840 L (222 gal)

Confirm that 840 L is larger than the minimum of 400 L

STEP 3: Infiltration (Figure 8.12)
Assessing Hydraulic Loading Rate (HLR) and calculating Linear Loading Rate (LLR)

As already mentioned, there are two methods for dispersing fluid into the landscape: 1) apply it over a specific area, such as a mulch basin; or 2) apply it through a linear piped infiltration system over some distance, as with a trench. Hydraulic Loading Rate

(HLR) is the rate wastewater enters the soil over an *area* (like over a mulch basin); it is measured in L/day/m² or gallons/day/ft². The Linear Loading Rate (LLR) is the rate of effluent applied per unit of length of the infiltration system; it is measured in L/day/m or gallons/day/ft. The more absorptive the soil, the smaller the area or shorter the distance of infiltration system needed. Figure 8.13 summarizes the differences between the two methods.

Fig. 8.15 (page 147): *Procedures to determine Daily Design Flow (DDF), the Adjusted Flow Volume (AFV), and Adjusted Tank Size (ATS) to be used in designing leachate systems.* ILLUSTRATION CREDIT: GORD BAIRD

Fig. 8.16 (below): *HLR table based on percolation rate, adapted from BC Ministry of Health Sewerage System Standard Practices Manual.* CREDIT: RALSTON AND PAYNE, 2014

STEP 3a: Finding the Area of Infiltrative Surface (AIS) dispersal using Hydraulic Loading Rate (HLR)

Knowing the HLR requires knowledge of the soil. We need to know the percolation rate, which is the time it takes for water to drop 1", as laid out in a perc test. The simple procedure for performing a percolation test is found in Appendix A.

For this example, we'll assume a soil perc rate of 3"/minute. We use this to look at the HLR table, found in Figure 8.16, and match up our perc rate to the HLR. Then the HLR is adjusted by multiplying it with the HLR Adjustment Factor found in Figure 8.15

column 5, providing us the *Adjusted HLR*, or AHLR.

Example:

Perc Rate = 3"/minute

HLR = 35 L/day/m²

AHLR = 35 L/day/m² × 0.7

AHLR = 24.5 L/day/m² or 0.5 gal/day/ft²

We finally complete the calculation using the Adjusted Flow Volume (AFV) from step 3 to calculate the *Area of Infiltrative Surface (AIS)* required for leachate.

AIS = AFV ÷ AHLR

AIS = 11.9 L/day ÷ 24.5 L/day/m²

AIS = 0.49 m² or 5.27 ft²

Therefore, if we chose to disperse the leachate into a mulch basin, the base area of the mulch basin would be required to be 0.5 meters square (5.27 ft²), coming from a dedicated tank sized at 840 L (222 gallons) to treat it to the degree required for dispersal.

STEP 3b: Finding the System Contour Length (SCL) using the Linear Loading Rate (LLR)

Similar to above, we begin by gathering soil data and percolation rate; again, we will stick with a rate of 3" per minute. Then we look at the Linear Loading Rate table (Figure 8.17) and match up our perc rate, the slope of our pipe, and the depth of absorbent soil under the trench. For our example, we will say that we are running at 3% slope and have 25 cm (10 in) soil below the infiltration surface.

Example:

Perc Rate = 3"/minute

LLR = 45 L/day/m

ALLR = 45 L/day/m × 0.7

ALLR = 31.5 L/day/m or 2.54 gal/day/ft

Percolation Test Range (min/inch)	Permeability KFS Range (mm/day)	Type 1 Effluent Loading Rate	
		L/day/m²	gal/day/ft²
>0.5	>17,000	45	0.91
0.5–1.0	8,000–17,000	45	0.91
1–2	4,000–8,000	45	0.91
2–4	2,000–4,000	35	0.71
4–7.5	1,000–2,000	30	0.61
7.5–15	550–1,000	27	0.55
15–30	300–550	23	0.47
30–60	150–300	15	0.30
60–120	75–150	12	0.24
>120	<75	Not appropriate	

| | | Step 1 | Step 2 | Step 3 | |
Category	Description	DDF Flow Volume Adjustment Factor	DDF Tank Volume Adjustment Factor	HLR Adjustment Factor	Other Notes
Daily Design Flow — Occupancy: Occupancy is derived from using the **# of bedrooms x 1.30 = # of occupants** (round up) **Daily Design Flow — Flow Volume:** DDF = # of occupants x per capita flow **Daily Design Adjusted Flow Volume:** AFV= DDF x flow volume adjustment factor **Daily Design Flow Adjusted Tank Volume:** ATV = (DDF x 3) x tank volume adjustment factor (and then further adjust sizing using notes)					
Combined Sewage	All combined wastewaters (blackwater and greywater)	1	3 x DDF volume	1	
Urine	Urinals and clean diverted urine at source	0.0085	2 x adjusted DDF volume	0.7 (when diluted 10:1)	Septic tank treatment is not necessary for dispersal
Leachate	Fluids collected from composting processors	0.0085	0.2 (min. tank size 400L)	0.7	
Blackwater	Combined flush toilet & urinal water	0.33	.08	1	
Brownwater	Flush toilet with urine diverted	0.33	.08	1	
Water separated from blackwater	Water separated from blackwater or brownwater stream.	.033	3 x adjusted DDF volume (min. tank size 1000L)	1	
Dark greywater (DGW)	Dishwasher and kitchen sink, bar sinks, mop sinks	0.07	0.2 (min. tank size 400L) (increase size 3 x if a garburator is used)	0.45	
Light greywater (LGW)	Bathtubs, showers, bathroom sinks, and laundry	0.65	0.4 (min. tank size 800L)	1.5	This value can be applied to size of a mulch basin — if no tank is used
Very light greywater (VLGW)	Bathtubs, showers, bathroom sinks.	0.65			
Laundry water	Laundry sink and washing machine	0.35			
Combined greywater	All wastewater **excluding** flush toilet and urine (no blackwater)	0.75	0.4 (min. tank size 800L) (increase size 2 x if a garburator is used)	1	
Blackwater + DGW or Brownwater + DGW	Combined streams include combining categories with one another for various streams of water flow	.04	0.9 (increase size 1.5 x if a garburator is used)	1	
LGW, laundry or combined + urine or leachate		Calculate the individual flows then add together to arrive at the combined flow volume	0.4 plus add 2 x daily urine flow or 2 x daily leachate flow and 2 x any flush water (min. tank size 800L) (increase size 2 x if a garburator is used)	1.2	
CGW + urine or leachate or LGW + leachate				0.75	
LGW + urine				1.2	This value can be applied to size of a mulch basin — if no tank is used

Again, using the Adjusted Flow Volume (AFV) from step 3 we can calculate the minimum System Contour Length (SCL) required for the leachate.

SCL = AFV ÷ ALLR

SCL = 11.9 L/day ÷ 31.5 L/day/m

SCL = 0.38 m (1.1 ft)

Therefore, if we chose to disperse leachate into an infiltration trench, the minimum length of infiltration pipe and trench would be 0.38 meters (1.1 ft).

To summarize, drainage/dispersal design may seem complex. This discussion really scratches the surface of a discipline that requires expertise in soil texture, soil structure, clay mineralogy, bulk density, effluent quality, soil aeration, percolation, and methods of application. Creating an integrated system incorporating combined flows is a task perhaps best left to a sewerage system designer. We hope that reading this has given you the basics about occupancy, flow, and storage volumes, adjusting those volumes for leachate (and other effluents), and how to size the dispersal system. The examples given are transferable to the other fluids listed in Figure 8.15. If you don't follow the steps given here, you will have an improperly designed system that can lead to failure; in particular, a poorly designed system will lead to clogging the surface of the soil, impeding infiltration (percolation), and reducing the loading rate of the soil.

Other Options for Leachate

There are other ways of addressing leachate without a "leachate drainage system." Leachate can be collected in containers similar to urine collection systems. Due to fecal contaminants, it will require a longer storage

Fig. 8.17: *LLR table based on percolation rate, adapted from BC Ministry of Health Sewerage System Standard Practices Manual.*
CREDIT: RALSTON AND PAYNE, 2014

LLR Rate is (L/day/m)		SLOPE 0-5%				SLOPE 5 TO < 10%				SLOPE 10 TO < 15%				SLOPE ≥ 15%			
Percolation Test Range (min/inch)	Permeability KFS Range (mm/day)	(CM) SOIL DEPTH BELOW INFILTRATIVE SURFACE				(CM) SOIL DEPTH BELOW INFILTRATIVE SURFACE				(CM) SOIL DEPTH BELOW INFILTRATIVE SURFACE				(CM) SOIL DEPTH BELOW INFILTRATIVE SURFACE			
		25–45	45–60	60–90	≥ 90	25–45	45–60	60–90	≥ 90	25–45	45–60	60–90	≥ 90	25–45	45–60	60–90	≥ 90
> 0.5	> 17,000	150	260	340	400	290	400	400	400	400	400	400	400	400	400	400	400
0.5–1.0	8000–17,000	85	140	180	250	150	250	330	400	260	400	400	400	360	400	400	400
1–2	4000–8000	50	80	110	140	80	140	180	250	140	240	310	400	200	330	400	400
2–4	2000–4000	45	55	70	90	55	85	110	150	90	140	180	240	120	190	240	340
4–7.5	1000–2000	45	55	70	70	50	60	75	90	60	90	110	140	80	120	150	200
7.5–15	550–1000	45	55	70	70	50	60	75	75	60	75	90	100	75	90	110	130
15–30	300–550	40	45	55	55	45	50	55	60	50	55	60	65	60	65	75	80
30–60	150–300	30	35	40	40	35	40	45	45	35	45	50	55	45	50	60	65
60–120	75–150	25	30	35	35	25	35	40	40	30	35	40	40	35	45	50	50
>120	<75	Not Applicable															

LLR Rate is (IG/day/ft)		SLOPE 0-5%				SLOPE 5 TO < 10%				SLOPE 10 TO < 15%				SLOPE ≥ 15%			
Percolation Test Range (min/inch)	Permeability KFS Range (mm/day)	(IN) SOIL DEPTH BELOW INFILTRATIVE SURFACE				(IN) SOIL DEPTH BELOW INFILTRATIVE SURFACE				(IN) SOIL DEPTH BELOW INFILTRATIVE SURFACE				(IN) SOIL DEPTH BELOW INFILTRATIVE SURFACE			
		10–18	18–24	24–36	≥ 36	10–18	18–24	24–36	≥ 36	10–18	18–24	24–36	≥ 36	10–18	18–24	24–36	≥ 36
> 0.5	> 17,000	10.1	17.5	22.8	26.9	19.5	26.9	26.9	26.9	26.9	26.9	26.9	26.9	26.9	26.9	26.9	26.9
0.5–1.0	8000–17,000	5.7	9.4	12.1	16.8	10.1	16.8	22.2	26.9	17.5	26.9	26.9	26.9	24.2	26.9	26.9	26.9
1–2	4000–8000	3.4	5.4	7.4	9.4	5.4	9.4	12.1	16.8	9.4	16.1	20.8	26.9	13.5	22.2	26.9	26.9
2–4	2000–4000	3.1	3.7	4.7	6.1	3.7	5.7	7.4	10.1	6.1	9.4	12.1	16.1	8.1	12.8	16.1	22.8
4–7.5	1000–2000	3.1	3.7	4.7	4.7	3.4	4.1	5.1	6.1	4.1	6.1	7.4	9.4	5.4	8.1	10.1	13.5
7.5–15	550–1000	3.1	3.7	4.7	4.7	3.4	4.1	5.1	5.1	4.1	5.1	6.1	6.8	5.1	6.1	7.4	8.8
15–30	300–550	2.7	3.1	3.7	3.7	3.1	3.4	3.7	4.1	3.4	3.7	4.1	4.4	4.1	4.4	5.1	5.4
30–60	150–300	2.1	2.4	2.7	2.7	2.4	2.7	3.1	3.1	2.4	3.1	3.4	3.7	3.1	3.4	4.1	4.4
60–120	75–150	1.7	2.1	2.4	2.4	1.7	2.4	2.7	2.7	2.1	2.4	2.7	2.7	2.4	3.1	3.4	3.4
>120	<75	Not Applicable															

time to sanitize it (more than six months). It can also be added fresh to compost processing piles (if the piles have a moisture content below 70%), providing beneficial moisture and nitrogen to promote thermophilic activity.

Two different approaches. The first is far more complex in the initial design and build but requires no further involvement in managing fluids. The second is simple, but requires regular monitoring and involvement in servicing. You now have the information to make the informed choice.

Evaporation: Fans

Fans are primarily intended to control for odor and humidity in both the pedestal and the chamber. In commode batch systems, fans are unlikely to control for anything more than just humidity and odor from the buckets. In moldering systems that utilize containers, fans are unlikely to impact moisture to any great degree as most of the surface area of the collected materials is encapsulated in the container, leaving only a small surface area open to evaporation.

In larger systems there is a wide surface area, and leachate drains through materials, so fans can play a big role in drying a pile. As noted, some systems may actually require the addition of fluids to stop the upper layer of the collected materials from drying out, and some, like the Phoenix, even have sprayers. On these larger surface systems, it is wise to use fans that have variable speed controls because they can give you some measure of control over the rate of drying.

Chapter 3 covers the pertinent information on fans.

Evaporation: Heaters

Heater elements are used to raise the temperature of collected materials. Keeping temperatures above 10°C (50°F) avoids bacteria from going dormant, but it is not a replacement for thermophilic natural heat. Generally, the more important role of heaters is to help evaporate moisture so that the fans can move it away. Most systems requiring heaters employ thermostats, so if moisture levels seem to be increasing, the temperature is adjusted higher to increase evaporation. Off-grid homes and cabins are not likely to employ these due to their energy demands. Heating moisture to evaporate it and exhaust it is extremely inefficient.

Small self-contained systems are more dependent on heaters. It's our opinion that larger compost systems requiring heat should be situated in a heated space. If temperature conditions are not conducive to running a compost system without heat, we advise using a compost toilet system that does not rely on warmth for evaporation, instead choosing one with a good drainage design, or one that utilizes outdoor composting processors.

References for greywater:

Ralston, I. "Manual of Composting Toilets." July, 2016.

Ludwig, A. "The New Create an Oasis with Greywater." 2016.

Gross, A. et al. *Greywater Reuse.* CRC Press. 2015.

References for urine dispersal:

von Münch and Winker. "Review of Urine Diversion." 2011.

Udert, Larsen, and Gujer. "Biologically Induced Precipitation." 2003.

References for dried urine storage:

Senecal and Vinnerås. "Urea Stabilisation." 2017.

Chapter 9

Best Practices

Testing and Records

In Chapter 2 you were introduced to the stages of raw, stabilized, cured (matured), and sanitized. These stages are based on time, temperature, and available nutrients. Keeping records is a great way to help you make informed choices on handling materials.

For some, the concept of testing or keeping records may sound crazy. But some jurisdictions *require* a homeowner to keep various records: a basic log of weekly temperatures; when materials were transferred from one stage of processing to another; the period of time a pile or moldering bin (or urine storage collection) has aged; and/or if any testing has been done to ensure that all damaging nitrogen forms have been converted. Even if this is not a requirement, you may wish to test your compost and keep records for the following reasons:

1. It will allow others to participate in the servicing by knowing the status of each part of the system.

2. It provides a pattern of servicing that new owners can use when homeownership changes.

3. It reduces the impacts resulting from "brain farts" and forgetful moments.

Solvita Compost Maturity Index Test

Matured and cured compost can be tested. Outside of lab testing for the values as expressed in Figure 2.19, or keeping records of *time and temperature* as presented in Figure 2.16, or relying on somewhat subjective qualities of homogeneity of particle size, earthy aroma, and observation of a pile shrinking 60%–90% from its original size, there is another test to determine the cured/matured state which can be performed at your compost pile: the Solvita test.

Solvita (the company) makes a variety of tests, one of which is called the Solvita Compost Maturity Index (CMI). It is a kit that is used on site: compost is added to a jar that has two probes, one that measures NH_3 (ammonia) and the other CO_2 (carbon dioxide). After several hours, the probes' colors are compared to a chart to determine the level of maturity. No training required.

Underlying principle of the test:

- CO_2 respiration: Levels of bacterial activity result in changing levels of CO_2; and, as microbial activity declines due to maturation, the CO_2 levels also decrease.

- Ammonia (NH_3): Volatile ammonia is converted to nitrate (a stable form) due to biological activity; therefore, volatile NH_3 declines during composting.

- Comparisons of NH_3 and CO_2: Provides a picture of how much bacterial activity has occurred and where the materials are in that process.

The test gives a result on a scale of 1–8 with "1" being raw, "5" being active and "8" being finished. Across North America and Europe, the Solvita test is the specified test required under most of the compost guidelines and regulations. A Solvita test score of 5–8 assures maturity has been achieved and on-site burial can occur.

Think of making a good compost like making cheese or wine. Knowing what you did and when you did it will really help you consistently achieve the end product you desire.

A basic kit costs about $200, and it provides materials to perform six tests. Materials can be ordered online direct from Solvita, solvita.com/store. Other sources would be local labs that provide soil testing; they may have kits in stock and for sale.

Summary of Best Practices

- Determine what your goals and needs are.
- Avoid systems that don't fit your lifestyle.
- Avoid systems that you can't guarantee resources for (electricity, water, bulking agent).
- Think about the future homeowners.
- Think about the future — about resilience.
- Think about the *what if's* (flooding, power outage, mechanical breakdowns).

Handling

- Think about the F Diagram (Figure 2.15) and use good judgement, use protective gloves, eyewear when needed, don't apply composted organics if they are not ready, think about keeping things out of your compost that could spread pathogens, and remember time and temperature for sanitization.
- Design like you plan to be 80 years old with no help — or who will help you?

Secured and Safe Processing

- Design compost bins to be aesthetically pleasing because, after all, we don't need to fuel others' prejudices.
- Keep the storage area (compost, tools, bins) safe, secure, and sanitary.
- Proper signage is essential.
- Have a maturation and/or sanitization plan including:
  - Know how much material you expect your household to produce.

- Know what the weekly tasks are.
  - List the monthly tasks.
  - List the yearly tasks.
- Know the storage times you need for the level of treatment desired:
  - Commode Batch = time and temperature at thermophilic and curing stages; dates
  - Moldering = bin labeling, time, temperature and dates in bin and other processors
  - Continuous = additional treatment (stored, composted), testing

Discharge

- Have a discharge plan.
- If not sanitizing, then meet the criteria in Chapter 2 and bury in a trench.
- If sanitizing, then ensure your logbook demonstrates time and temperatures in your plan are minimally met. Consider testing.
- Don't apply to food crops unless you meet criteria set out in Figures 2.18 and 2.19 (regarding mineral content and pathogen/stability).

Recordkeeping and Logbook

The following is an example of the headings and information (how and what) you would keep for your records.

Primary Purpose:

The primary purpose of this system is ...

Example answer: To collect and process liquids and solids to create an ecologically safe material to recycle nutrients back onto the landscape.

Site:

Write down all the information regarding your site. I know I might try to skip this, saying I know it all ... for myself. This is not just for your own use (though it may help you think through many issues), it is also for someone coming after you. It may also be handy information to have on hand for any health inspectors/regulators.

- Location/address.
- Climatic zone.
- Seasonal precipitation patterns (this influences when and how sanitized materials are discharged).
- Soils, slopes, wetlands, setbacks, potential for flood risk.
- Site Map:
  - Distance to property lines, water, well.
  - Location of compound (secured storage and/or composters).
  - Location of fruit trees etc. for surface discharge.
  - Location of site burial locations, those areas where you have buried materials that were not processed well enough for surface discharge.

System Design:

Write a brief one-sentence overview of the system. Then include details about use (what kind and how much bulking agent, daily cleaning, how to use, etc.).

Details:

- Number of occupants served, for what time frame (full time/recreational).
- Type of pedestal (direct drop, foam-flush, mechanical, jet, etc.)
- What are the specific handling stages?
  - initial collection (bucket, bin, chamber, vault)
  - additional storage or transferring for processing (transfer to moldering bin or compost bin)
- Time targets for servicing:
  - daily/weekly/bi-weekly servicing

  - monthly
  - yearly
- Details about storage area (how it is secured, what signage, layout, dedicated tools, how protected from excess precipitation)
- Moisture handling:
  - urine diversion/storage/usage
  - leachate diversion
  - other (e.g. Aquatron filtering)

Curing and Sanitizing:

- What are the required TIME and TEMPERATURES at the various processing stages?
- What is the curing pathway for the designed system?
- What is the sanitization pathway for the designed system?
- What testing or documentation do you have? (Untested, but with good documentation; Solvita test; lab testing; and a list of standards to be tested to).

Discharge:

- At what level of process is it discharged (cured and or sanitized)?
- How is it discharged (site burial in trench or surface)?
- When? (What time of year is soil above 10°C (50°F); when is best time to avoid excess precipitation?)
- Where will discharge occur?

Logbook:

Keeping a log is essential. It's so easy to skip recordkeeping. Don't assume you'll be able to remember dates and temperatures. The information you log will be different for different systems (see Figure 9.1). If you skip this step, you virtually guarantee failure.

Moldering

Date: Empty bin install	Date: Bin filled	Date: Bin emptied	Does it meet minimum time/temp? Yes or No		If no... Date: Of additional step	Date: Discharge	Total time in months from filled to discharge	Date: Tested	Additional notes, discharge location, etc.
			Cured?	Sanitized?					

Carousel

Date: Empty chamber start	Date: Rotated to new chamber	Total # months from empty to full	Date: Chamber is emptied	Total # months of moldering	Does this meet sanitization plan? Yes/No	If no... Date of additional step	Date: Finished processing	Date: Tested	Additional notes, additional steps, discharge location, etc.

Continuous   Date: Start up/install

Date: Inspection of raw materials chamber	Date: Inspection of finished materials chamber	Date: Finished materials removed to batch processing	Required months in batch processing to be cured or sanitized (as per plan)	Date: Discharged	Date: Tested	Additional notes, discharge location, etc.

Commode Batch

Date: Start of active compost pile	Date & temperature: Addition to active pile		Date & temperature: LAST addition to active pile		Date: Expected to meet sanitization	Date: Harvested for discharge	Date: Tested	Additional notes, discharge location, etc.
	Date	Temp. °	Date	Temp. °				

Maintenance Log

Insert list of regular maintenance to be followed (e.g. Service fans yearly, check leachate drains monthly, clean drop chute, etc.)		Scheduled? Yes/No	Additional notes and comments
Date	Description of action (what and how)		

Fig. 9.1: *Sample logbook fields for different systems.* Illustration credit: Gord Baird

Chapter 10

The Last Flush

WESTERN CULTURE'S CURRENT CONCEPT of sanitation is not nearly so elegant or evolved as the biological complexities of soil. Our sewer systems rely on manufactured infrastructure to collect and treat water, relying heavily on chlorine and other chemicals to sterilize and make it potable — while, ironically, the manufacture of these chemicals poisons us, with chlorine being the primary producer of dioxins in our environment. Our sewer systems require pipes and treatment plants to remove contaminants from the once-potable water, and then we have to treat the remaining sludge. This takes an enormous amount of energy and materials, all paid for with our tax dollars. Additionally, all of this infrastructure depreciates as it wears out, ultimately ending up in landfills — at further cost to society.

We have discussed using compost toilets as a way to avoid these negatives, but what if there were limited water, no power, no bulking agents, and no space to compost? If we look outside of the suburbs of Western cities into inner cities or places on the planet where populations are very dense, we see that millions of people have no water, no sewer, no power, and not even a small space for a compost pile. Presently, 2.6 billion people don't have access to safe, sanitary toilets; annually, 280,000 deaths are directly tied to lack of sanitation (WHO, "Sanitation," 2017). In many places, women and girls put themselves at great risk of sexual assault when using public toilets; they often resort to waiting until dark to venture into the streets to defecate.

Can we not solve these problems? After all, we have used science and engineering to create air travel, artificial intelligence, power storage, and water purification; we've split the atom, and we can live in space. How is it that we have not applied that level of human ingenuity to creating appropriate toilet technology? Bill Gates's answer is this: those who have privilege and the access to resources and knowledge "weren't getting sick and dying from contaminated water, and they didn't know anyone who was" (Gates, 2015).

In 2011, the Bill and Melinda Gates Foundation initiated the "Reinvent the Toilet Challenge" (RTTC), offering grants totaling $3.4 million to those who dare look at human manures from a whole new perspective. The RTTC asked for the invention of a toilet that would disinfect human resources within 24 hours, be inexpensive to build and operate ($0.05 per person per day), require no water hook-up, no power hook-up, or a sewerage or leachate system, AND one that can transform the resources into energy, clean water and fertilizer. WOW...now, there's a challenge.

Here's where science is coming close to matching the overarching functions and objectives of compost toilets to create solutions wherein human resources are just that, resources. The Gates Foundation narrowed down to several technologies that met their criteria: the Blue Diversion Autarky toilet from EAWAG/EOOS (Swiss Federal Institute of Aquatic Science and Technology); the Sol-Char from the University of Colorado, Boulder; and the Nano Membrane Toilet developed at Cranfield University.

Why discuss a high-tech toilet? In the first chapter, we introduced the reasons why someone would choose a compost toilet. Throughout the following chapters you learned about different technologies that would answer those criteria. These new high-tech toilets meet all those same needs, for the same reasons, with the same outcomes, but (eventually) they will do it at a low cost. Here's an introduction to just one of these inventions — which represents the level of research that is occurring across all these technologies.

Cranfield University Nano Membrane Toilet

Research on the development of the Cranfield Toilet began in September 2012, and continues full steam ahead with a group of 24 full-time engineers and the support of another 20 staff. This team is developing a bio-reactor that converts solids to fertilizer, a membrane technology that converts urine to fresh water and fertilizer, a design that enhances social and cultural integration of the toilet, and a business model that supports and services it. Research is publicly documented and shared regularly in journals like *Environmental Science, Journal of Membrane Science, Energy Conversion and Management,* and *Fuel.*

Highlights:

- The Nano Membrane Toilet is a self-sustained system.
- Up to 87% of the total amount of water fed to the system can be recovered for use.
- Human excrement is loaded with energy that can be recovered — between 23.1 and 69.2 watt hours/kg of settled solids.

- Net power output of the entire system is similar to a USB port at peak power (so, 2–6 watts). This means one could charge a cell phone from a toilet. Yes! Seriously!
- Dry human feces has a heating value of 24 MJ/kg and can handle continued sustained combustion despite the 75% moisture content found within feces.
- More energy can be recovered from pyrolysis (burning) of feces than converting it to biogas fuel.

The system has several basic components, which are a bit different from conventional compost toilet components, as you might imagine. There is a fairly conventional seat, which is a pleasant user interface, you might say. The action of lifting the toilet seat up ratchets a collection cup toward the user; this cup also acts as a seal between the toilet bowl and the chamber below. When one has completed their business and puts the lid down, the ratcheting mechanics tip the contents into the chamber and scrapes the cup with something akin to a big spatula, readying it for the next user. Inside the chamber, the solids sink through the fluids and end up at a special inclined "screw" that raises the solids out of the chamber, and, in the process, both drains and compresses them, readying them to be deposited into a self-powered micro-incinerator that produces heat and energy to power the toilet. The fluids overflow into a system which is heated from the reactor; they then pass through a nano-membrane that condenses the steam into purified water that is collected and stored for reuse. The remnants of the incinerated feces are in the form of sterilized ash that can be used for soil conditioning.

Each unit is designed to handle household daily usage for ten people; they are stand-alone, self-contained, and unobtrusive.

This system is potentially a disruptive technology that may change the way we design our biggest cities.

The team wanted to design a product that would be acceptable for societies that are accustomed to the modern flush toilet (despite the technology being so old). If it is acceptable and functional in all places, cultures, economies, climates … then we have a technology that is disruptive.

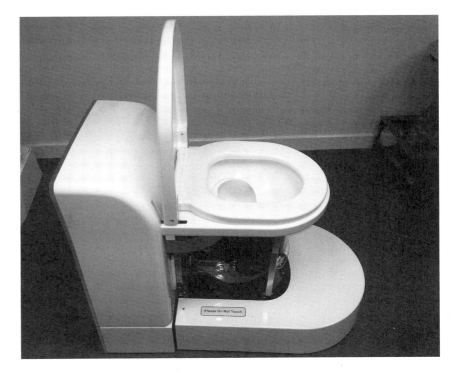

Fig. 10.1: *Cranfield Nano Membrane Toilet prototype, winner of the Reinvent the Toilet Challenge, incorporated social, business, cultural, energy, water, and agricultural research into the development of a self-contained resource-collection toilet.* Photo credit: Alison Parker, Cranfield Water Science Institute, Cranfield University, 2017

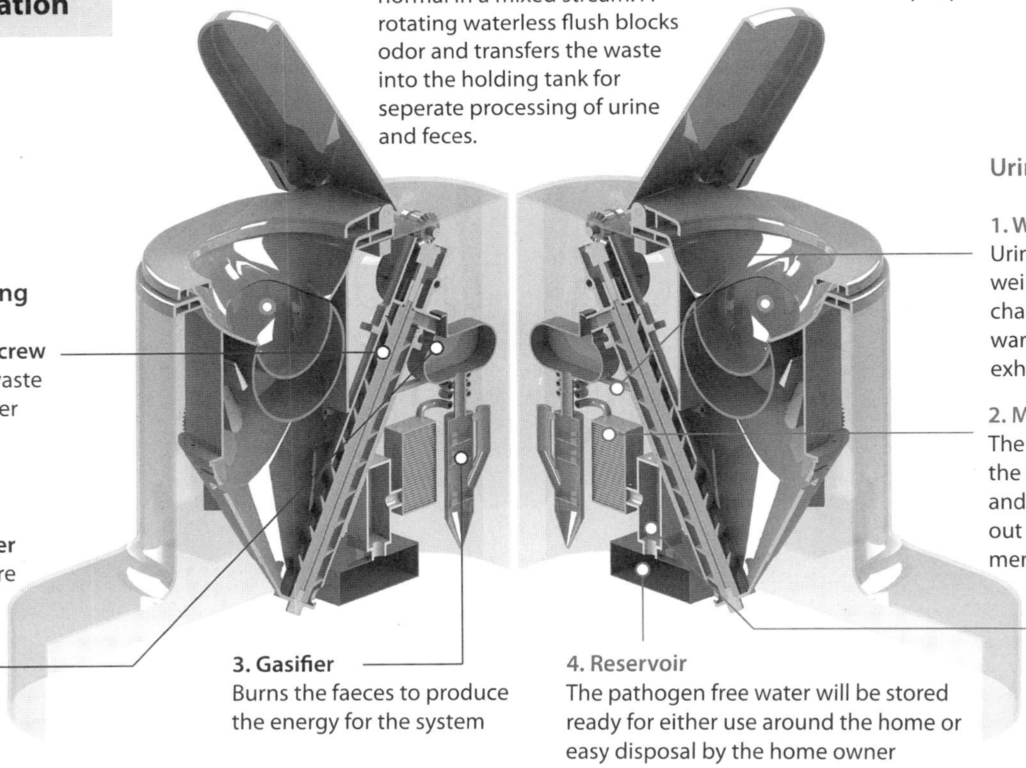

System Configuration

Rotating flush
Waste enters the toilet as normal in a mixed stream. A rotating waterless flush blocks odor and transfers the waste into the holding tank for seperate processing of urine and feces.

A waterless self-contained toilet for private household of 10 people

Feces processing

1. Archimedes screw
Removes solid waste from holding after settling period

2. Drier pelletizer
Reduces moisture content of the solid waste before dosing the fuel into the gasifier below

3. Gasifier
Burns the faeces to produce the energy for the system

4. Reservoir
The pathogen free water will be stored ready for either use around the home or easy disposal by the home owner

Urine processing

1. Weir channel
Urine will pass over the weir and into the channel where it will warm up around the exhaust of the gasifier

2. Membrane bundle
The urine will pass into the membrane chamber and pure water will pass out of the hollow membrane fibres

3. Heat exchanger
The water vapour will condense to liquid and fall to the bottom

Fig. 10.2: *Inside the Cranfield Nano Membrane Toilet there are components that allow for a paradigm shift in resource recovery to occur.* Illustration credit: Ross Tierney, Cranfield University

Conclusion

We have had thousands of people tour our Eco-Sense home and learn about all the systems. Without exception, we are always amazed by the reactions and take enjoyment when they inevitably hit that point, when they become aware of the connection between natural ecological systems and the human constructed world. People are always very surprised. They view our humanure composting piles, greywater, and worm bio-filters and say: "Wow! What a simple system!"

Our response: "Yes," and "No, not really."

Yes, it is simple when we stand back and think that natural "simple" biological systems do most of the work. Then again, not really, when we more closely observe the complexity of millions of years of co-evolution of microorganisms and all their symbiotic partnerships performing complex functions — so detailed that we simply cannot manufacture the same system. The thing is, why would we even try to mimic these systems using plastic and metal when nature has already perfected all of this in the soil? The answer to that is presented above. Even the Cranfield Toilet, as engineered as it may be, is focused on simplifying what it takes to manage wastes by removing the need for massive complex infrastructure and creating simple resources that do not pollute. In permaculture we call this "stacking functions." We can meet our needs with simple systems and the use of appropriate technology, resulting in ecologically sustainable and resilient systems. We know what perfection looks like… it looks like living soil. Our challenge is to design our engineered systems for full resource recovery using nature as our architect.

French writer and poet Antoine de Saint-Exupéry wisely noted, "Perfection is achieved not when there is nothing more to add, but when there is nothing left to take away." Compost toilets systems, especially the very simple commode batch or humanure system, are the ultimate design. There is nothing left to take away.

Appendix A: Percolation Test Procedure

Introduction

The following procedure was adapted from the Province of British Columbia's *Sewerage System Standard Practice Manual,* Volume 3 (2014) (Ralston & Payne, 2014), and is presented in its entirety as written in the Rainwater Management Standards for the City of Victoria (Public Works and Engineering, 2015).

Percolation Test Procedure

Use the following instructions to conduct a percolation test. This is the procedure for a percolation test in BC.

Step 1:
Test holes should be 30 cm (12 in) square or 36 cm (14 in) round and excavated to the proposed depth of the management method. It is generally easiest to dig a larger hole part way down, then dig a 18 to 20 cm (7 to 8 in) deep accurately sized test hole in the base of the larger hole.

Step 2:
To make the percolation test more accurate, any smeared soil should be removed from the walls of the test holes. This is best achieved by digging the hole approximately 5 cm under-sized (2 in) and then enlarging the hole to the accurate size as follows: using a rigid knife, insert the blade into the top side of the hole opposite you approximately 2.5 cm (1 in) deep, holding the blade with its cutting edge vertical. Pull the blade away to break out a chunk of soil, repeat about an inch (2.5 cm) apart around the hole, then repeat for another "ring" below until reaching the base. The result will be a hole with a ragged inner surface which looks like a freshly broken clod of soil.

Step 3:
The base of the hole should be cleaned of debris and be approximately flat. To achieve this, use a metal scoop or similar. It should also be picked to present a natural surface. Note that a picking action (use a pointed tool) is needed, not a scratching action (which just produces smears that are indented).

Step 4:
Place 5 cm (2 in) of clean fine gravel in the bottom of the hole. If the sidewalls are likely to collapse, use a paper basket to support the sidewalls (see below). Place a piece of white plastic (or something similar) provided with clear marks at 5" and 6" from the bottom of the test hole prior to adding the gravel. For greater accuracy, a float and pointer arrangement can be set up.

Step 5:
If the soil contains considerable amounts of silt or clay, and certainly for any soil with "clay" as part of the texture description, the test holes should be pre-soaked before proceeding with the test. Pre-soaking is accomplished by keeping the hole filled with water for 4 hours or more. The water should be added carefully and slowly to avoid disturbing the soil (including the sidewall soils). The test should be carried out immediately after pre-soaking.

Step 6:
To undertake the test, fill the test hole (the accurately sized test hole) with water. The water should be added carefully and slowly to avoid disturbing the soil (including the sidewall soils). When the water level is 5" or less from the bottom of the hole, refill the hole to the top. No recording of time needs be done for these two fillings.

Step 7:

When the water level after the second filling (procedure [6]) is 5" or less from the bottom of the hole, add enough water to bring the depth of water to 6" or slightly more. Note that these measurements are from the base of the soil bottom (using the marker installed in step [4]), not the gravel layer.

Step 8:

Observe the water level until it drops to the 6" depth, at precisely 6", commence timing, when the water level reaches the 5" depth, stop timing, record the time in minutes.

Step 9:

Repeat procedures (7) and (8) until the last two rates of fall do not vary more than 2 minutes per inch or by more than 10% (whichever is less).

Step 10:

Report slowest rate for each hole.

Step 11:

Backfill the holes with the excavated soil and flag and label their locations so you can mark them up for any site map or plan.

If a test hole is discarded due to flow in a root channel or similar, record the information and make a replacement test. If there is a large variation (greater than or equal to 50%) between tests in the same soil layer, increase the number of tests.

Paper Basket to Protect Hole

If sidewalls of the hole are likely to collapse, one option is to make a paper basket to protect and support the sidewalls as follows:

- Cut the bottom out of a large paper bag (grocery bag) and cut the bag open along a side.
- Lay bag on a soft surface. Punch holes in the bag about 5 to 7.5 cm (2 to 3 in) apart using a pencil or something similar.
- Roll into a tube, with the short dimension being the axis of the tube, and place it in the test hole.
- Open the tube until the paper is in contact with the sidewalls of the test hole, then roll the top of the tube over to stiffen it.
- After placing the tube in the hole, place the plastic marker and add the base gravel layer.

Percolation rate for design

Location (address):							File #	
Date:		Tested by:						
Weather:								
Test number	Depth of base of hole from surface (cm)	Timings, minutes per inch for water to drop from 6" to 5" from base hole.						Lowest rate (minutes per inch)
		#1	#2	#3	#4	#5	#6	
1								
2								
3								
4								
5								
6								
7								
8								
							Percolation rate:	
Notes:								

Appendix B: Manufacturers

Chapter 5: Batch Systems

Loveable Loo — Humanure Store
143 Forest Lane
Grove City, PA 16127
Phone: 814-786-9085
Email: mail@josephjenkins.com
Web: humanurehandbook.com/store/

Chapter 6: Moldering Systems

Separett Canada:
Canadian Eco Products
53 Lantern St.
Kitchener, ON
Phone: 855-402-9735
Email: separett@rogers.com
Web: separett.ca

Separett USA:
Separett-USA/EcoServices
PO Box 226
Barrington, NH 03825
Phone: 800-682-8619
Email: info@separett-usa.com
Web: separett-usa.com

Separett Sweden:
Separett AB
Bredastensvägen 8
331 44 Värnamo, Sweden
Telephone: +46 (0)370-307200
Fax: +46 (0)370-307236
Email: info@separett.com

ECOJOHN
Basic Waterless Toilet and Incinerating
Toilets
6148 Getty Dr.
North Little Rock, AR 72117
Phone: 866-ECOJOHN (326-5646)
Web: ecojohn.com/ecojohn_basic.html

Ekolet Finland:
Ekolet Ltd.
Estetie 3
00430 Helsinki, Finland
Phone: +358106662690
Email: info@ekolet.com
Web: ekolet.com

Aquatron International AB
Ekebyvägen 4
725 92 Västerås, Sweden
Phone: +46 10 490 10 40
Email: info@aquatron.se

NOTE: North American distributor was in
transition at time of writing. Website will
have links to updated distributors.

Ecodomeo Head Office
20 allée Pierre Louis Guilliny
La pépinière
ZA Les Laurons, 26110 Nyons, France
Phone: +33 (0)6 86 15 13 66
Web: ecodomeo.com

Ecodomeo USA & Canada :
Toilet Tech Solutions
4220 24th Ave W,
Seattle, WA
Phone (CA): 604-505-3656
Phone (US): 206-713-7805
Web: toilettech.com

BioLet Toilet Systems, Inc.
830 West State Street
Newcomerstown, OH 43832
Email: info@biolet.com
Web: biolet.com/products/biolet-
composting-toielet-ne

EcoTech Products
508 Boston Post Road
Weston, MA 02493
Phone: 855-864-5980
Email: info@ecotechproducts.net
Web: ecotechproducts.net

Biorealis Systems, Inc.
P.O. Box 77-2773
Eagle River, AK 99577
Website: biorealis.com
Web (plans): biorealis.com/composter/
rotating/

Chapter 7: Continuous Systems

MullToa (also known as BioLet and EcoLet)
Canada
EcoEthic Inc.
Robert Davis
23 Thompson Road
Sunderland, ON L0C 1H0
Phone: 888-436-3996
Fax: 705-357-9971
Email: ecoinfo@ecoethic.ca
Web: ecoethic.ca

USA
BioLet Toilet Systems
830 West State Street
Newcomerstown, OH 43832
Phone: 800-524-6538
Email: info@biolet.com
Web: biolet.com

Phoenix Toilet USA
Advanced Composting Systems
195 Meadows Rd.
Whitefish, MT 59937
Phone: 406-862-3854
Fax: 406-862-3855
Email: phoenix@compostingtoilet.com

Phoenix Toilet Canada
Sunergy Systems LTD
Box 70
Cremona, Alberta T0M 0R0
Phone: 403-637-3973
Fax: 403-637-3973
Email: sunergy@compostingtoilet.com

Glossary

A

Additives: Bulking agents, or other amendments added to composting toilet pedestals or processors.

Aerobic: Relating to, involving, or requiring free oxygen; or growing or occurring only in the presence of molecular oxygen, as in "aerobic bacteria" or "aerobic composting."

Aerobic (composting process): Having an oxygen content in all parts of the compost over 5%.

Anaerobic: Characterized by the absence of oxygen, or growing in the absence of molecular oxygen (as in "anaerobic digestion").

B

Batch process: A composting process wherein decomposition or composting occurs without further addition of raw materials.

Blackwater: Source-separated wastewater from flush toilets.

Brownwater: Source-separated feces, toilet paper, and water from flush toilets excluding urine (urine diverted).

Bulking agent: Material added to collected excreta for the purpose of improving structure, creating porosity, absorbing liquid, controlling odor and, in some cases, adding carbon.

C

Chamber batch collection system: A collection system for excreta or wastewater which involves collecting the material in a container and physically transporting it to a separate processor (treatment system).

Combined sewage wastewater system: A wastewater system receiving all wastewater flows from a building.

Commode: A composting toilet pedestal with container collection system.

Compost: Organic matter that is a stable, safe, mature product resulting from composting. Composting toilet residual organic matter is *not* compost.

Composting: Managed process of bio-oxidation of organic matter of heterogeneous organic substrate including a thermophilic phase of specific minimum time and minimum temperature followed by a curing step.

Compost toilet: A composting toilet system that includes collection, conveyance, storage, and processing of waste to a stable, safe state through either decomposition or composting.

Continuous process: A composting processor continually accepts new materials while allowing one or more steps of the composting process to occur, ultimately providing decomposed "finished" materials.

Curing: The final stage of decomposition or composting, characterized by a resting period at ambient temperature in order to reduce volatile nitrogen and other phytotoxins, also called aging, or maturing.

D

Daily design flow (DDF): Estimated peak ground discharge flow of wastewater, used in designing/sizing wastewater systems.

Dark greywater: Greywater flowing from kitchen, dishwashing, and mop sink uses.

Diversion (urine diversion): Separation of urine at source. (Not to be considered separation.)

E

Excreta: Feces, toilet paper, and urine.

F

Finished material: Residual organic matter or solid waste end product that is removed from a composting toilet processor. Residual organic matter is *not* compost.

G

Greywater: Water bearing waste from the preparation of food and drink, dishwashing, bathing, showering, and general household cleaning and laundry. Greywater includes subcategories of dark greywater, light greywater, and very light greywater.

Greywater diversion system: A sanitary drainage system to allow diversion of one or more greywater streams from the combined wastewater flow.

Greywater seasonal diversion system: A sanitary system allowing for seasonal diversion of greywater for seasonal irrigation use.

H

Headspace (container): Additional storage volume or free space over and above minimum required volume of a container.

L

Leachate: Effluent discharged from a composting toilet or composting toilet processor.

Light greywater: Greywater flowing from uses other than kitchen, dishwashing, or mop sinks. Example: shower water or laundry greywater.

M

Maturation: See "curing."

Maturity: The agricultural or soil amendment value or impact of the residual organic matter. Represented by reduction of phytotoxic compounds and ammonia. Maturation occurs after initial digestion and stabilization. Maturity is related to *stability*, but also represents the impact of compost chemical properties on plant development and soil organisms.

Mesophilic (temperature): Of the temperature range 20–40°C (68°F–104°F); used to describe the phases, and organisms present, at this temperature range.

Molder (moulder): Cool-temperature aerobic decomposition, characterized by mesophilic or psychrophilic temperatures resulting in slow decay and decomposition.

Mulch basin: A shallow subsurface irrigation basin or bed with mulch media covering the area of infiltrative surface (area of native ground that accepts moisture).

Multiple barrier (multi-barrier) risk management: A barrier is a part of the treatment or handling chain of an on-site sewerage system which either reduces pathogen levels in excreta, residual material or effluent, or which reduces the risk of human exposure to pathogens.

O

On-site burial (composting toilet residual organic matter): Subsurface discharge of residual organic matter by burial, preferably in the form of a trench rather than hole or pit.

On-site surface discharge (composting toilet residual organic matter): Surface discharge (including incorporation to the surface soil) of residual organic matter.

P

Psychrophilic (temperature): Of the temperature range −10°C–20°C (14°F–68°F); used to describe the phases, and organisms present, at this temperature range. Dominant temperature for moldering composting toilet processors and during late stage curing.

S

Source separation: Separation of wastes into specific types of material at the point of generation.

Source-separated wastewater: Wastewater stream that has been discharged from the building by a separate sanitary drainage or collection system, or has been separated from solids. Includes greywater, diverted urine, water separated from blackwater and leachate. Includes diversion and separated greywater sanitary drainage systems.

Stability: The first stage of decomposition or composting; characterized by low or no odor and decreased phototoxicity to plants, though highly active biological processes are indicative of early stages of decomposition.

Struvite: Magnesium ammonium phosphate (MAP), which precipitates from urine.

T

Thermophilic (temperature): Of the temperature range of 45°C–80°C (113°F–176°F) used to describe the phases, and organisms present, at this temperature range. This is the required temperature to define true composting.

V

Vector: A route of transmission of pathogens from source to victim; can include insects, dogs, birds, rodents, and humans.

Very light greywater: Wastewater from showers, baths, hand basins. Light greywater excluding laundry water.

Volatile solids (organic matter): Components (primarily carbon, oxygen, and nitrogen) that can be burned off.

References

(*NOTE:* Access dates appear in square brackets.)

Advanced Composting Systems. January 1997. *Design Features of the Phoenix.* [November 1, 2017] www.composting toilet.com/LITRACK/design.pdf

Advanced Composting Systems: Phoenix Composting Toilets. n.d. "The Phoenix in Residences and Cabins." [November 1, 2017] www.compostingtoilet. com/?page_id=401

Advantage Environment. January 24, 2017. *Natural Filter Media for Phosphorus Recovery.* [November 24, 2017] advantage-environment.com/workplace/natural-filter-media-for-phosphorus-recovery/

Alberts, B., et al. 2002. "Introduction to Pathogens," in *Molecular Biology of the Cell* (4th ed.) Garland Science. www.ncbi.nlm. nih.gov/books/NBK26917/

Aquatron International AB. October 2011. *Aquatron Separator: Installation and Maintenance User's Manual.* [November 24, 2017] www.aquatron.se

Balch, G.C., et al. 2013 *Phosphorous Removal Performance of Bioretention Soil Mix Amended with Imbrium Systems Sorbtive Media.* (p 37) Centre for Alternative Wastewater Treatment and Fleming College. www.imbriumsystems.com

Barnhart, E., and H. Maingay. January 2015. *A New, User Friendly, Urine-Diverting, Waterless, Composting Eco-Toilet.* The Green Center Inc. capecodecotoilet center.files.wordpress.com

Betancourt, W.Q., and L.M. Shulman. 2015. "Polioviruses and Other Enteroviruses." In J.B. Rose and B. Jiménez-Cisneros (eds.), *Global Water Pathogens Project.*

Lansing, Michigan: UNESCO. Retrieved from www.waterpathogens.org/book/polioviruses-and-other-enteroviruses

Bohl, E.H. 1952. "Epidemiological Studies on Leptospirosis." (Doctorate dissertation). Ohio State University. Retrieved from etd.ohiolink.edu/!etd.send_file?accession=osu14863978 41219839&disposition=inline

Buswell, C.M., et al. 1998. "Extended Survival and Persistence of *Campylobacter* spp. in Water and Aquatic Biofilms and Their Detection by Immunofluorescent-Antibody and -rRNA Staining." *Applied and Environmental Microbiology* 64(2), 733–741.

Calloway, D.H., and S. Margen. 1971. "Variation in Endogenous Nitrogen Excretion and Dietary Nitrogen Utilization as Determinants of Human Protein Requirement." *The Journal of Nutrition* 101(2), 205–216.

Canada Mortgage and Housing Corporation. 2002. *Dual-flush Toilet Testing.* [June 2, 2017] www.cmhc-schl.gc.ca/publications/en/rh-pr/tech/02-124-e.html

Canadian Council of Ministers of the Environment, and Compost Guidelines Task Group. 2005. *Guidelines for Compost Quality.* Canadian Council of Ministers of the Environment. www.ccme.ca/files/Resources/waste/organics/compost gdlns_1340_e.pdf

Carballa, M., et al. 2004. "Behavior of Pharmaceuticals, Cosmetics and Hormones in a Sewage Treatment Plant." *Water Research* 38(12), 2918–2926. www.sciencedirect.com/science/article/pii/S0043135404001642?via%3Dihub

Carballa, M., F. Omil, and J.M. Lema. 2008. "Comparison of Predicted and Measured Concentrations of Selected Pharmaceuticals, Fragrances and Hormones in Spanish Sewage." *Chemosphere* 72(8), 1118–1123. doi. org/10.1016/j.chemosphere.2008.04.034

City of Victoria Stormwater Utility. June 2015. *Rainwater Management Standards.* (Professional Edition). www.victoria.ca/assets/Departments/ Engineering~Public~Works/ Documents/SWVictoria_Professional_ Rainwater_Mgmt_Stds_June2015.pdf

Clivus Multrum Australia. n.d. *Maintaining your Clivus Multrum.* [November 1, 2017] www.clivusmultrum.com.au/ maintaining-your-clivus-multrum

Clivus New England. n.d. *How the Clivus Multrum Works.* [November 1, 2017] www.clivusne.com/science-and technology.php

Cordell, D., and S. White. 2014. "Life's Bottleneck: Sustaining the World's Phosphorus for a Food Secure Future." *Annual Review of Environment and Resources* 39(1), 161–188. www.annualreviews.org/doi/10.1146/ annurev-environ-010213-113300

Crosby, R.L. n.d. *A Low Cost Site-Built Composting Toilet System.* Biorealis Systems, Inc. [November 24, 2017] biorealis.com/composter/rotating/

Czemiel, J. 2000. "Phosphorus and Nitrogen in Sanitary Systems in Kalmar." *Urban Water* 2(1), 63–69. www.science direct. com/science/article/pii/S1462075800 000406?via%3Dihub

Danielsson, M., and M. Lippincott. n.d. "A Sewer Catastrophe Companion: Dry Toilets for Wet Disasters." Portland Bureau of Emergency Management.

www.portlandoregon.gov/pbem/ article/447707

Del Porto, D., and C. Steinfeld. 2000. *The Composting Toilet Systems Book: A Practical Guide to Choosing, Planning and Maintaining Composting Toilet Systems — A Water-Saving, Pollution-Preventing Alternative.* (1st ed.) The Center for Ecological Pollution Prevention.

W.B. DeOreo, P. Mayer, B. Dziegielewski, and J. Kiefer. April 2016. *Residential End Uses of Water, Version 2: Executive Report.* www.waterrf.org/PublicReportLibrary/ 4309A.pdf

Department of Economic and Social Affairs, United Nations. 2017. *World Population Prospects: The 2017 Revision: Key Findings and Advance Tables* (No. ESA/P/WP/ 248). esa.un.org

Eastwood, M.A. 1973. "Vegetable Fibre: Its Physical Properties." Proceedings of the Nutrition Society, 32(3), 137–143. doi.org/10.1079/PNS19730031

Ecoflo Wastewater Management. n.d. "How to maintain your Clivus Multrum tank?" YouTube. [November 1, 2017] www.youtube.com/embed/H9Iwg9v Aaxs

ECOJOHN Portable Incinerating Toilets. n.d. [November 24, 2017] ecojohn.com/ ecojohn_basic.html

ENVIS Centre on Hygiene, Sanitation, Sewage Treatment Systems and Technology. October 2016. *Pathogens in Human Excreta.* sulabhenvis.nic.in/Database/ sanitation_humanexcreta_2131.aspx

Feachem, R.G., et al. 1981. *Appropriate Technology for Water Supply and Sanitation: Health Aspects of Excreta and Sullage Management: A State-of-the-Art Review.* (No. 11508). World Bank Studies in Water Supply and Sanitation.

documents.worldbank.org/curated/en/ 929641467989573003/pdf/ 11508000 Approp0te0of0the0art0review.pdf

Feachem, R.G., et al. 1983. *Sanitation and Disease: Health Aspects of Excreta and Wastewater Management* (PUB No. 11616). World Bank Studies in Water Supply and Sanitation. documents. worldbank.org/curated/en/704041468 740420118/pdf/multi0page.pdf

Full Circle Composting Toilets. n.d. [October 19, 2017] fullcirclecompost.org

Gates, B. August 11 2015. *Update: What Ever Happened to the Machine That Turns Feces into Water?* [November 10, 2017] www.gatesnotes.com/Development/ Omni-Processor-Update

Germer, J., et al. 2010. "Temperature and Deactivation of Microbial Faecal Indicators During Small Scale Co-Composting of Faecal Matter." *Waste Management* 30(2), 185–191. doi. org/10.1016/j.wasman.2009.09.030

Goldblith, S.A., and E.L. Wick. 1961. *Analysis of Human Fecal Components and Study of Methods for Their Recovery in Space Systems.* (No. ASO-TR-61–419). NASA. contrails.iit.edu/files/original/ ASDTR61-419.pdf

Gotaas, Harold Benedict and the World Health Organization. 1956. "Composting: Sanitary Disposal and Reclamation of Organic Wastes." World Health Organization. www.who.int/iris/ handle/10665/41665

Gross, A., et al. 2015. *Greywater Reuse.* CRC Press.

Hanak, D.P., et al. 2016. "Conceptual Energy and Water Recovery System for Self-Sustained Nano Membrane Toilet." *Energy Conversion and Management* 126, 352–361. www.sciencedirect.com/

science/article/pii/S01968904163066 28?via%3Dihub

Haug, R.T. 1993. *The Practical Handbook of Compost Engineering.* Lewis Publishers.

Hill, G.B., and S.A. Baldwin. 2012. "Vermicomposting Toilets, an Alternative to Latrine Style Microbial Composting Toilets, Prove Far Superior in Mass Reduction, Pathogen Destruction, Compost Quality, and Operational Cost." *Waste Management* 32(10), 1811–1820. doi.org/10.1016/j.wasman.2012. 04.023

Hill, G.B., S.A. Baldwin, and C. Lalander. 2013. "The Effectiveness and Safety of Vermi- Versus Conventional Composting of Human Feces with *Ascaris suum* Ova as Model Helminthic Parasites." *Journal of Sustainable Development* 6(4).

Jenkins, J. 2005. *The Humanure Handbook: A Guide to Composting Human Manure* (3rd ed.) Joseph Jenkins, Inc.

Jennings, M.C., K.P.C. Minbiole, and W.M. Wuest. 2015. "Quaternary Ammonium Compounds: An Antimicrobial Mainstay and Platform for Innovation to Address Bacterial Resistance." *ACS Infectious Diseases* 1(7), 288–303. pubs.acs.org/ doi/10.1021/acsinfecdis.5b00047

Jensen, P.K., et al. 2009. "Survival of Ascaris Eggs and Hygienic Quality of Human Excreta in Vietnamese Composting Latrines." *Environmental Health* 8, 57. www.ncbi.nlm.nih.gov/ pubmed/20003550

Katayama, V. 2015, January 14. "Norovirus and Other Caliciviruses." [December 3, 2017] www.waterpathogens.org/book/ norovirus-and-other-caliciviruses

Land, B. 2003. *Sweet Smelling Toilet Installation Guide* Recreation Management Tech Tips No. 0323

1303-SDTDC. United States Department of Agriculture, Forest Service. www.fs.fed.us/eng/pubs/pdf/03231303.pdf

Larsen, T.A., and J. Lienert. 2007. *No Mix: A New Approach to Urban Water Management.* Eawag Aquatic Research. www.novaquatis.eawag.ch/publikationen/final_report_E.pdf

Lienert, J., and T.A. Larsen. n.d. "Novaquatis: Practical Guide." [June 8, 2017] www.novaquatis.eawag.ch/ueberblick/praxis/index_EN.html

Lippincott, M., et. al. 2015. "Moving Beyond the NSF: Composting Toilet Systems." Recode: Legalizing Sustainability. legislature. vermont.gov/assets/Documents/ 2016/WorkGroups/House%20Fish%20and%20Wildlife/Bills/H.375/Witness%20Testimony/H.375~Kai%20Mikkel%20Forlie~Moving%20Beyond%20the%20NSF-Composting%20Toilet%20Systems~4-28-2015.pdf

Ludwig, A. 2016. *The New Create an Oasis with Greywater: Integrated Design for Water Conservation: Reuse, Rainwater Harvesting & Sustainable Landscaping.* (6th ed.) Oasis Design.

Meinzinger, F., and M. Oldenburg. 2009. "Characteristics of Source-Separated Household Wastewater Flows: A Statistical Assessment." *Water Science and Technology: A Journal of the International Association on Water Pollution Research* 59(9), 1785–1791. wst.iwaponline.com/content/59/9/1785

Nelson, G., and M. Kerfoot. February 2006. "Phoenix Composting Toilet System Instructions for Operation and Maintenance." Advanced Composting Systems. www.compostingtoilet.com/LITRACK/manual.pdf

Nilsson, C., et al. 2013. "Efficacy of Reactive Mineral-Based Sorbents for Phosphate, Bacteria, Nitrogen and TOC Removal: Column Experiment in Recirculation Batch Mode." *Water Research* 47(14), 5165–5175. www.sciencedirect.com/science/article/pii/S0043135413004880?via%3Dihub

NSF International. n.d. "NSF/ANSI 41: Non-Liquid Systems," [October 19, 2017] www.nsf.org/services/by-industry/water-wastewater/onsite-wastewater/non-liquid-saturated-treatment-systems

Olson, M. 2018. *Human and Animal Pathogens in Manure.* University of Calgary.

Onabanjo, T., et al. 2016. "An Experimental Investigation of the Combustion Performance of Human Faeces." *Fuel* 184, 780–791. www.sciencedirect.com/science/article/pii/S0016236116306809?via%3Dihub

Onabanjo, T., et al. 2016. "Energy Recovery from Human Feces Via Gasification: A Thermodynamic Equilibrium Modelling Approach." *Energy Conversion and Management* 118, 364–376. www.sciencedirect.com/science/article/pii/S019689041630245X?via%3Dihub

Potworowski, J. A. March 2010. *The Transformation of the National Building Code of Canada: From Prescriptions to Objectives.* Tefler School of Management, University of Ottawa. sites.telfer.uottawa.ca/makingithappen/files/2014/05/Building-Code-Case-Study.pdf

Ralston, I. July 2016. *Manual of Composting Toilets and Greywater Practice.* (E. Hoeppner, et al., eds.). BC Ministry of Health, Health Protection Branch. www2.gov.bc.ca/assets/gov/environment/

waste-management/sewage/provincial-provincial-composting-toilet-manual.pdf

Ralston, I., and M. Payne. 2014. *Sewerage System Standard Practice Manual.* (Ver. 3). BC Ministry of Health, Health Protection Branch. www2.gov.bc.ca/assets/gov/environment/waste-management/sewage/spmv3-24september2014.pdf

Raven, P.H., and G.B. Johnson. 2002. "Extremophilic Bacteria and Microbial Diversity," in *Biology* (6th edition) McGraw-Hill. www.mhhe.com/biosci/genbio/raven6b/graphics/raven06b/enhancementchapters/raven30_enhancement.html

Richard, T., and N.M. Trautmann. n.d. "C/N Ratio," Cornell Composting. [August 19, 2017] compost.css.cornell.edu/calc/cn_ratio.html

Rieck, C., E. von Münch, and H. Hoffmann. 2013. "Technology Review of Urine Diverting Dry Toilets (UDDTs): Overview of Design, Operation, Management and Costs." Sustainable Sanitation Project. Deutsche Gesellschaft für Internationale Zusammenarbeit. Retrieved from www.susana.org/_resources/documents/default/2-874-technology-review-of-uddts-18-june-2013.pdf

Rose, C., et al. 2015. "The Characterization of Feces and Urine: A Review of the Literature to Inform Advanced Treatment Technology." *Critical Reviews in Environmental Science and Technology* 45(17), 1827–1879. www.tandfonline.com/doi/full/10.1080/10643389.2014.1000761

Schlussler, L. April 2002. "Energy Efficient Shower Conserves Water and Also Reduces Condensation Without Ventilation." [June 5, 2017]

www.sunfrost. com/energy_efficient_shower.html

Senecal, J., and B. Vinnerås. 2017. "Urea Stabilisation and Concentration for Urine-Diverting Dry Toilets: Urine Dehydration in Ash." *Science of the Total Environment* 586, 650–657.

Schouw, N. L., et al. 2002. "Composition of Human Excreta: A Case Study from Southern Thailand." *Science of the Total Environment* 286(1–3), 155–166.

Schröder, P., et al. 2016. "Status of Hormones and Painkillers in Wastewater Effluents Across Several European States: Considerations for the EU Watch List Concerning Estradiols and Diclofenac." *Environmental Science and Pollution Research International* 23, 12835–12866. doi.org/10.1007/s11356-016-6503-x

Silvester, K.R., et al. 1997. "Effect of Meat and Resistant Starch on Fecal Excretion of Apparent N- Nitroso Compounds and Ammonia from the Human Large Bowel." *Nutrition and Cancer* 29(1), 13–23.

Simha, P., and M. Ganesapillai. 2016. *Ecological Sanitation and Nutrient Recovery from Human Urine: How Far Have We Come? A Review* (27) www.sciencedirect.com/science/article/pii/S246820391630173X?via%3Dihub

Sobrados-Bernardos, L., and J.E. Smith. 2012. "Controlling Pathogens and Stabilizing Sludge/Biosolids: A Global Perspective of Where We Are Today and Where We Need To Go." *Proceedings of the Water Environment Federation* 20122, 56–70. static1.squarespace.com/static/54806478e4b0dc44e1698e88/t/548612e2e4b027feceb088f6/1418072802448/Smith-ControllingPathogens-GlobalPerspective-2013.pdf

Sossou, S.K., et al. 2014. "Inactivation Mechanisms of Pathogenic Bacteria in Several Matrixes during the Composting Process in a Composting Toilet." *Environmental Technology* 35(5–8), 674–680. doi.org/10.1080/09593330.2013.841268

Spuhler, D. 2018, April 27. "Urine Storage." [May 16, 2018] www.sswm.info/water-nutrient-cycle/wastewater-treatment/hardwares/site-storage-and-treatments/urine-storage

Stanford University. n.d. "Transmission of *Hymenolepis* spp." [October 28, 2017] web.stanford.edu/group/parasites/ParaSites 2002/hymenolepsis/transmission.htm

Stanford University/ParaSites. *Index of Group Parasites.* [January 9, 2018] web.stanford.edu/group/parasites/

Stanford University/ParaSites 2012. *The Mentor Initiative: Devoted to Reducing Malarial Deaths and Suffering in Humanitarian Crises.* [January 9, 2018] web.stanford.edu/group/parasites/ParaSites2012/

Stauffer, B. n.d. *Vacuum Toilet.* SSWM. [November 16, 2017] www.sswm.info/water-nutrient-cycle/wastewater-treatment/hardwares/user-interface/vacuum-toilet.

Ternes, T., and A. Joss (eds.). 2008. *Human Pharmaceuticals, Hormones and Fragrances: The Challenge of Micropollutants in Urban Water Management.* (Reprinted) IWA Publ.

Toilet Revolution. July 30 2016. "Urine Diverting or Not?" [June 8, 2017] www.toiletrevolution.com/news/urine-diverting/

Trautmann, N.M., T. Richard, and M.E. Krasny. n.d. "Compost Chemistry."

Cornell Composting. [June 11, 2017] compost.css.cornell.edu/chemistry.html

Udert, K.M., T.A. Larsen, and W. Gujer. 2003. "Biologically Induced Precipitation in Urine-Collecting Systems." *Water Science and Technology: Water Supply* 3(3), 71–78.

Vinnerås, B. January 2002. *Possibilities for Sustainable Nutrient Recycling by Faecal Separation Combined with Urine Diversion.* Swedish University of Agricultural Sciences, Uppsala. www.researchgate.net/publication/ 35245723_Possibilities_for_sustainable_nutrient_recycling_by_faecal_separation_combined_with_urine_diversion

Vinnerås, B., et al. 2006. "The Characteristics of Household Wastewater and Biodegradable Solid Waste: A Proposal for New Swedish Design Values." *Urban Water Journal* 3(1), 3–11. www.tandfonline.com/doi/abs/10.1080/15730620600578629

von Münch, E., and M.I. Winker, 2011. *Technology Review of Urine Diversion Components.* Eschborn, Germany: Deutsche Gesellschaft für Internationale Zusammenarbeit (GIZ) GmbH. www.susana.org/_resources/documents/default/2-875-giz2011-en-technology-review-urine-diversion.pdf

Wang, C.Y., et al. 2017. "Tube-Side Mass Transfer for Hollow Fibre Membrane Contactors Operated in the Low Graetz Range." *Journal of Membrane Science* 523 (Supplement C), 235–246. www.sciencedirect.com/science/article/pii/S0376738816305750?via%3Dihub

Webber, R. 2016. *Communicable Diseases: A Global Perspective* (5th ed). CABI.

WHO. 2006. *Guidelines for the Safe Use of Wastewater, Excreta and Greywater*

(3rd ed.) vol. 4, World Health Organization.

WHO. "Sanitation." July, 2017. [December 30, 2017] www.who.int/mediacentre/factsheets/fs392/en/

Wichuk, K., and D. McCartney. 2007. "A Review of the Effectiveness of Current Time-Temperature Regulations on Pathogen Inactivation During Composting." *Journal of Environmental Engineering and Science* 6, 573–586. 10.1139/S07-011.

Wichuk, K., and D. McCartney, 2010. "Compost Stability and Maturity Evaluation: A Literature Review." *Canadian Journal of Civil Engineering* 37, 1505–1523.

Winblad, U., and M. Simpson-Hebert (eds.). 2004. *Ecological Sanitation.* Stockholm Environment Institute.

Worm Farm Facts. n.d. "Red Worms." [October 22, 2017] www.wormfarmfacts.com/Red-Worms.html

Wuebbles, D.J., et al. 2017. *Executive Summary: Climate Science Special Report: Fourth National Climate Assessment.* U.S. Global Climate Research Program. science2017.globalchange.gov

Ylösjoki, M. n.d. "Composting Toilets for Home: Indoor and Outdoor Dry Composting Toilet Systems." [November 24, 2017] ekolet.com

Index

Page numbers in *italics* indicate figures.

About the Authors

The authors, Ann and Gord with their dog Nina, sit in front of their solar powered cob home surrounded by gardens.

ANN BAIRD, BSc AND GORD BAIRD, BA are the owners and co-creators of the internationally recognized Eco-Sense home in Victoria, BC, Canada. The building of the Eco-Sense home occurred at a time when greywater, rain water harvesting, compost toilets, and earthen architecture were on the fringes of cultural acceptance and not yet supported by local regulations. By challenging the codes and regulations in a logical, informed, and respectful manner, their home became the first legal, seismically engineered, two-story load-bearing cob home in North America, with approval eventually being given to all the systems mentioned. Additionally, the Eco-Sense home was the first Living Building Challenge (LBC) project to be audited. The LBC is the most challenging green building rating system globally, and their home was the first to achieve petal recognition leading to the status of "World's Greenest Modern House" for a number of years.

The Baird's insatiable appetite to learn and effect change through both policy and cultural story, have spawned a broad range of opportunities to work on many aspects of different projects. They have given countless tours, courses, and presentations for groups, including engineers, designers, regulators, bankers, politicians, government staff, trades and technical schools, grade school students, local food producers, and garden groups. In 2014, Gord was asked to participate as a technical editor on a team contracted by the BC Ministry of Health to write the *Manual of Composting Toilet & Greywater Practice* which became law in September 2016.

The Baird's integrated lifestyle weaves food, water, energy, and livelihood into a very resilient and rewarding life, which they call a 3-Thirds Lifestyle. The equal balance between work, passion, and volunteer activities creates a very rewarding and meaningful life. Their homestead has become a working farm with a perennial edible food forestry nursery that produces 100% of their fruits and vegetables. This living classroom provides for teaching and consulting on the topics of rainwater harvesting, greywater, compost toilets, food forestry, food and garden systems, building codes, sustainable energy and living roofs. They've engaged in research on energy, moisture, and thermodynamics for cob and earthen mass wall systems. Both have sat as elected municipal councilors (2014–2018), with the next term yet to be determined. Gord has acted as a Water Commissioner for both the CRD (Capital Regional District regional government) Water Commission, and Juan de Fuca Water Distribution Commission.

Gord is an ARCSA AP (American Rain Catchment Systems Association Accredited Professional) and has sat as a director on CANARM (Canadian Association for Rainwater Management).

Their home, lifestyle, and accomplishments have been extensively featured in the media including documentaries, books, magazines, TV news, radio, and even featured in a year-long museum exhibit of the Royal British Columbia Museum's Free Spirit exhibition.

Most important of all is that they have lived full time with compost toilets, (Gord since 2005 and Ann since 2000), and have designed and built toilets for both public and private use. The Bairds built a public compost toilet in their community, and for many years have had the contract to clean the bathroom and collect the waste (resource) for composting. They are intimately familiar with what works and what doesn't. For this reason, Gord and Ann were asked to author this book by New Society Publishers for the Sustainable Building Essentials Series.

Being asked is one thing, but being motivated is completely different. One of the primary motivations for writing this

book derived from an upsetting situation that unfolded with one of their clients. One day, unannounced, a government employee arrived at a client's house with a minivan full of child seats to remove the children from the home. A determination had been made by someone in one branch of government that using a compost toilet was akin to child abuse. This government employee asked no questions and demonstrated no understanding of policy and regulations existing in another branch of the same government. Consequently, a decision based on preconceived ideas — not information, existing policy, or science — was made. Needless to say, rapid work on several people's part quickly led to one side of the government educating the other side, and the kids were not removed. In this case, education was the critical solution.

Removing uninformed preconceptions about composting toilets using science, experience, and policy work, should go a long way to allow and even encourage site-appropriate ecological sanitation acceptance to grow. Their goal is to take compost toilets from the waste stream to the mainstream.

A Note About the Publisher

NEW SOCIETY PUBLISHERS is an activist, solutions-oriented publisher focused on publishing books for a world of change. Our books offer tips, tools, and insights from leading experts in sustainable building, homesteading, climate change, environment, conscientious commerce, renewable energy, and more — positive solutions for troubled times.

We're proud to hold to the highest environmental and social standards of any publisher in North America. This is why some of our books might cost a little more. We think it's worth it!

- We print all our books in North America, never overseas
- All our books are printed on **100% post-consumer recycled paper**, processed chlorine free, with low-VOC vegetable-based inks (since 2002)
- Our corporate structure is an innovative employee shareholder agreement, so we're one-third employee-owned (since 2015)
- We're carbon-neutral (since 2006)
- We're certified as a B Corporation (since 2016)

At New Society Publishers, we care deeply about *what* we publish — but also about how we do business.

Download our catalog at https://newsociety.com/Our-Catalog or for a printed copy please email info@newsocietypub.com or call 1-800-567-6772 ext 111

New Society Publishers

ENVIRONMENTAL BENEFITS STATEMENT

For every 5,000 books printed, New Society saves the following resources:[1]

43	Trees
3,854	Pounds of Solid Waste
4,240	Gallons of Water
5,531	Kilowatt Hours of Electricity
7,006	Pounds of Greenhouse Gases
30	Pounds of HAPs, VOCs, and AOX Combined
11	Cubic Yards of Landfill Space

[1]Environmental benefits are calculated based on research done by the Environmental Defense Fund and other members of the Paper Task Force who study the environmental impacts of the paper industry.